Ultrafast Biophotonics

More information about this series at http://www.springer.com/series/3740

BIOLOGICAL AND MEDICAL PHYSICS, BIOMEDICAL ENGINEERING

The fields of biological and medical physics and biomedical engineering are broad, multidisciplinary and dynamic. They lie at the crossroads of frontier research in physics, biology, chemistry, and medicine. The Biological and Medical Physics, Biomedical Engineering Series is intended to be comprehensive, covering a broad range of topics important to the study of the physical, chemical and biological sciences. Its goal is to provide scientists and engineers with textbooks, monographs, and reference works to address the growing need for information.

Books in the series emphasize established and emergent areas of science including molecular, membrane, and mathematical biophysics; photosynthetic energy harvesting and conversion; information processing; physical principles of genetics; sensory communications; automata networks, neural networks, and cellular automata. Equally important will be coverage of applied aspects of biological and medical physics and biomedical engineering such as molecular electronic components and devices, biosensors, medicine, imaging, physical principles of renewable energy production, advanced prostheses, and environmental control and engineering.

P. Vasa · D. Mathur

Ultrafast Biophotonics

Springer

P. Vasa
Department of Physics
Indian Institute of Technology Bombay
Mumbai
India

D. Mathur
Tata Institute of Fundamental Research
Mumbai
India

ISSN 1618-7210 ISSN 2197-5647 (electronic)
Biological and Medical Physics, Biomedical Engineering
ISBN 978-3-319-39612-5 ISBN 978-3-319-39614-9 (eBook)
DOI 10.1007/978-3-319-39614-9

Library of Congress Control Number: 2016942516

Printed on acid-free paper

This Springer imprint is published by Springer Nature
The registered company is Springer International Publishing AG Switzerland

Preface

This book discusses contemporary and emerging techniques of ultrafast science which are helping to open entirely new vistas for probing biological entities and processes. These include the use of femtosecond lasers to facilitate time-resolved imaging, multiphoton microscopy, single molecule studies, laser surgery and, also, to create plasma channels in aqueous media that help detect stress marker proteins and probe DNA damage induced by slow electrons and radicals. We review some of the topics that are presently on the horizon, like the use of coherent control, squeezed light, frequency combs, terahertz imaging, the possibility of mimicking biological systems and, perhaps surprisingly to many, invoking quantum mechanical effects such as coherent superposition, radical pair production and tunneling to rationalize phenomena like photosynthesis, avian navigation in the earth's magnetic field and respiration. Also discussed is the role played by ultrafast biophotonics in developing biomimetic devices whose quantum functionalities may be "engineered" for applications in light-harvesting, solar energy conversion, magnetic field sensing, photonic devices and single biomolecular electronics.

It is hoped that the interdisciplinary contents of this book will be of benefit to physical scientists and life scientists alike.

We would like to thank Dr. Claus Ascheron, Physics Editor of Springer-Verlag at Heidelberg for his enthusiastic support of this writing project.

Mumbai, India

P. Vasa
D. Mathur

Contents

About the Authors

P. Vasa is currently an Associate Professor in the Physics Department of the Indian Institute of Technology Bombay, Mumbai. Her research interests include experimental studies of ultrafast phenomena, nanooptics, and plasmonics. Amongst her awards and recognitions include membership of the Indian National Young Academy of Science, the Humboldt Fellowship, the Schlumberger Foundation's "Faculty for the Future" Award, the European Union's Heritage Fellowship, the Government of India's SERB Women Excellence Award and Indian National Science Academy Medal for Young Scientist.

D. Mathur is Distinguished Professor at Tata Institute of Fundamental Research, Mumbai. His research interests encompass ultrafast, high-intensity laser science, and diverse topics in the atomic, molecular and optical sciences, and in biological physics. Amongst his recognitions are the Fellowship of TWAS (the World Academy of Science), the Indian National Science Academy, and the Indian Academy of Sciences. He has been awarded the Government of India's S. S. Bhatnagar Prize, the J.C. Bose National Fellowship, the European Union's Erasmus-Mundus Scholarship in Optical Science and Technology, the Royal Society's Guest Research Fellowship, the Association of Commonwealth Universities' Fulton Fellowship, and various named Lectureships. He has been co-Editor of EPL (Europhysics Letters) and member of the Editorial Advisory Board of the Journal of Physics B and Rapid Communications in Mass Spectrometry.

Chapter 1
Introduction and Overview

Abstract This chapter presents an overview of the prospects of present and anticipated experimental and theoretical advances in ultrafast biophotonics discussed in this book. The areas include multiphoton and multidimensional microscopy, ultrafast single molecule studies, femtosecond laser surgery, quantum biology, biomimetic devices, tunneling in biological entities, cavitation in aqueous biological media by thermal as well as plasmonic means, coherent control of biochemical processes and probing biological entities beyond the quantum limit. Emerging techniques that are likely to play a major role in the development of the subject are also discussed; they include ultrafast multi-dimensional spectroscopy, generation of femtosecond-long X-ray pulses from free electron lasers for ultrashort imaging of biological materials without causing photodamage, application of terahertz radiation for imaging and spectroscopy, uses of frequency comb Fourier transform spectroscopy, and exciting advances in the emerging area of biomimetic technology.

What is photonics?

A succinct definition might be: *photonics is the science and technology of light, with emphasis on applications*. One of the distinguishing features of the science of photonics lies in the natural and apparently seamless linkages that the subject provides between fundamental scientific studies and technological applications in diverse areas of contemporary and futuristic importance. The extension of photonics to *biophotonics* is a rather obvious one: it concerns, on the one hand, the use of optical science to gain insights into biological phenomena and, on the other, utilizing optical science to drive the development of new biological and biomedical methodologies and technologies. The mutually beneficial dialectical relationships between experiment, theory, and applications that are a hallmark of biophotonics are serving to accelerate developments within the field. The prognosis for the near future is exciting!

As can be imagined, the canvas that becomes available on the basis of the above definition of biophotonics is immensely wide. This book focuses on only that subset of the canvas that deals with, or relies on, ultrafast phenomena and processes. The term *ultrafast* is used in reference to timescales that are prevalent in dynamics that

P. Vasa and D. Mathur, *Ultrafast Biophotonics*, Biological and Medical Physics, Biomedical Engineering, DOI 10.1007/978-3-319-39614-9_1

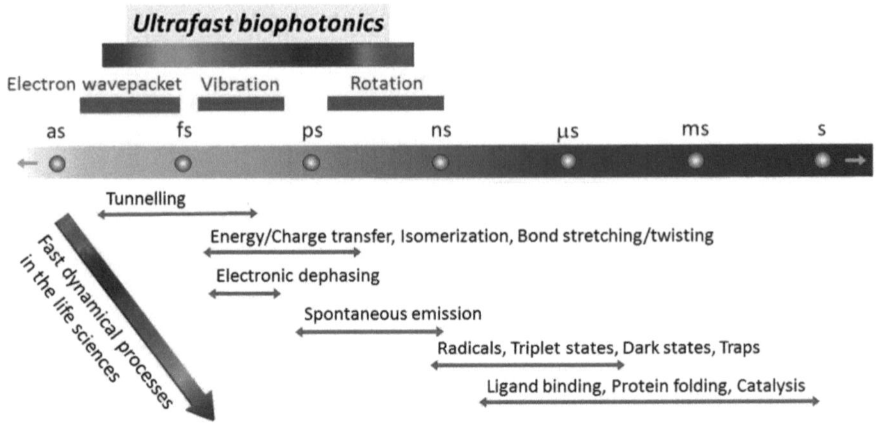

Fig. 1.1 Schematic depiction of some fast and ultrafast dynamical processes encountered in contemporary biology

occur within individual atoms and molecules, the constituents of all living matter. These timescales vary from tens and hundreds of picoseconds that single molecules take to rotate, to tens of femtoseconds that molecules take to vibrate to, finally, tens and hundreds of attoseconds—the time taken for electronic motion within atoms and molecules (Fig. 1.1). In the context of biophotonics, this book shall explore biological phenomena that occur on such short timescales and, also, show how photonics techniques based on ultrafast interactions can enable these phenomena to be probed.

It is not difficult to appreciate that tools, methods and concepts initially developed for diverse endeavours in physics in general, photonics in particular, are playing an increasingly important role in biology; indeed, some of them have become an accepted feature of many a contemporary life sciences laboratory. Raman and fluorescence spectroscopy, especially multidimensional and multiphoton varieties, are examples that come readily to mind. It is also not unexpected that biology itself is starting to stimulate physical scientists to develop afresh methods, tools, and concepts that, of necessity, induce close participation of life scientists: "super-resolution" microscopy, multiphoton imaging, and the computational modeling of single-molecule trajectories are amongst several contemporary examples.

In the course of this book recourse has been taken to discuss processes that appear to occur on picosecond, or even nanosecond, timescales. Whenever we refer to such "long" timescales, they will pertain to the composite of several sub-processes, each of which occur on ultrashort, usually femtosecond, timescales. A challenge that remains to be tackled is to experimentally "deconvolute" these ultrafast sub-processes from the currently observable process.

1.1 Living Systems

Living systems have been around on earth for at least 4 billion years. In order to distinguish living systems from inanimate matter, we articulate the following features: Living entities show

(i) the ability to self-reproduce,
(ii) the ability to harvest energy;

Furthermore,

(iii) they evolve, and
(iv) they manifest complexity.

Complexity is one distinguishing feature: even a single living cell comprises over 1000 different molecules. While complexity is also found in inanimate objects, it remains a fact that as far as the other three features are concerned, there are no analogues to be found in the physical sciences that possess such characteristics.

There are also similarities that should be articulated. As is known from several examples, the physical sciences are witness to length scales and time scales that span very many orders of magnitude: from time periods of billions of years since the primordial big bang to electron dynamics within atoms and molecules that occur on attosecond (10^{-18} s) time scales. Biological entities and processes also span many orders of magnitude. In terms of size and parameter space, life scientists readily deal with a single molecule of water on the one hand, and the effect on the ecosystem of the dynamic energy budget of the earth's atmosphere on the other.

It is known that a single living cell—such as that of *E. coli*—can transcribe as many as 5×10^6 genes in the course of $\sim$2000 s or less. This implies that the real-time information processing capabilities of a single simple cell are far in advance of any practical computational devices that physical scientists and engineers can envision at present. A single cell can process about 10 Gb of information per hour while occupying a volume of only a few μm^3! We note that there is nothing special about a cell of *E. coli*; it is a relatively simple cell, coded by only about five million base pairs. One can extrapolate to multi-cellular systems and it becomes very obvious that biological entities present very remarkable information capabilities, much in excess of what even the most advanced silicon-based non-living systems can hope for at present.

1.2 Energy Scales

Important entities that one encounters in the life sciences are DNA, RNA, and proteins. The covalent bonds that constitute the backbone of such polymers have evolved to be quite robust and, typically, their binding energies are $\Delta G \geq 1$ eV. However, forces that act in a direction that is perpendicular to the polymer backbones can be considerably weaker. But, surprisingly (from the perspective of a physical scientist), these weak forces often influence, in major fashion, the form, dynamics, and functions of many biomolecules. Hence, one frequently encounters forces in the life sciences that are very much smaller than ΔG—of the order of $k_B T$—and these tiny forces often prove to be important. Here, T denotes the temperature (either the laboratory temperature in the case of a process or the body temperature in the case of a living entity) and k_B is the Boltzmann constant.

In the context of energy scales, it is pertinent to compare the vastly different energy scales that are encountered throughout the life sciences compared to those that are routinely encountered in physics, chemistry and engineering. Molecular interactions that drive most biological processes work on k_BT energy scales. As $k_B N_A = R$, where R is the gas constant and N_A denotes Avogadro's number, 1 k_BT = 0.62 kcal/mole at a temperature of 300 K. It is also instructive to compare k_BT = 25 meV at room temperature with energy levels in the prototypical hydrogen atom, ~10 eV, and covalent bonds and binding energies of atoms in metals, ~1 eV. The comparison becomes even more stark on a macroscopic level: an object that would be considered tiny in the context of the physical sciences—one weighing, say, 1 mg—being displaced with a speed of 1 cm s^{-1} has an energy of 10^9 eV!

1.3 Overview of the Book

We begin the book by presenting, in Chap. 2, an overview of some basic concepts related to light, nonlinear optics, and ultrashort pulses on which many of our explorations of ultrafast biophotonics will be based. We introduce, in pedagogical fashion, concepts like the intensity of laser light, ponderomotive energy, nonlinear optics, frequency conversion; also introduced are methods of generating and characterizing ultrashort pulses, and how such pulses propagate through matter.

Structural studies have acted as significant precursors of many contemporary developments in the life sciences, with structure determination of biological entities often being the precursor to the discovery of tangible links to their biological function. Chapter 3 deals with a well-established use of ultrafast nonlinear optics for structure determination: biophotonic microscopy. The essential physics behind such microscopy is that of multiphoton excitation using pulses of intense infrared light which enable greater depth penetration to be achieved, an important consideration in microscopy. This chapter also discusses the successes of super-resolution microscopy, which is based essentially on the "manipulation" of molecular spectra, and the anticipated successes of 4D electron microscopy in unravelling ultrafast biological phenomena. Among the biophotonic applications discussed in this chapter is one that concerns protein fibrillization, specifically the potential involvement of amyloid fibrils in diseases like Parkinson's and Alzheimer's diseases which share fibril formation as the common symptom. Unfortunately, for long it has been the case that Alzheimer's disease (AD) could only be diagnosed upon analysis of postmortem tissue; such analysis invariably reveals the existence of extensive neurofibrillary tangles and plaques of beta-amalyoid (βA). Multiphoton microscopy offers in vivo imaging of βA plaques in intact brains, with much superior spatial resolution than competing methods like positron emission tomography, magnetic resonance imaging, and fluorescence imaging. Tangible benefits of superior spatial resolution include the ability to image objects like dendritic spines, the post-synaptic apparatus of excitatory synapses as well as the onset and subsequent growth of amyloid plaques, and Ca^{2+} transients. Second harmonic microscopy is more complex than multiphoton microscopy as account has to be taken of a coherent optical process that demands phase matching from all parts of the nonlinear focal volume. However, retention of phase information permits useful information to be obtained from signal directionality, a facet of particular importance in samples like type I collagen fibrils (whose diameters are almost the same as the wavelength of visible light). This has allowed in vivo studies of tissues under dynamical conditions like the healing of wounds, development, and malignancy. "Super-resolution" microscopy has allowed imaging of cultured hippocampal neurons from neonatal rats, with images revealing the existence of nanometre-sized protein clusters. 3D mapping of neurofilaments in differentiated neuroblastoma cells has also been accomplished!

In the last two decades, the detection and spectroscopy of individual molecules has begun to find widespread applications in the life sciences. While conventional time-resolved experiments continue to provide information about ensemble-averaged properties, ultrafast single molecule techniques are able to track the photodynamics of individual molecules, revealing their unique transient intermediates. Chapter 4 presents an overview of single molecule techniques like surface enhanced coherent anti-Stokes Raman scattering, Förster resonance energy transfer, pump-probe spectroscopy, and pulse-shaping; these can not only identify but also provide information about electronic/vibrational wavepacket interferences and relaxation mechanisms at the level of individual molecules. As an illustration of their relevance in ultrafast biophotonics, a study of persistence of coherence in an individual photosynthetic complex under physiological conditions is presented.

Chapter 5 reviews two other well-established applications of ultrafast biophotonics: femtosecond laser surgery and cell manipulation. The essential physics that drives these applications is discussed, including quantitative considerations of parameters that affect biophotonic applications. At the tissue level, ablation induced by ultrafast pulsed lasers has allowed the "cutting" of minute volumes of tissue and to facilitate transection of cells: sub-micrometre surgery of neuronal and vascular entities are now routinely performed as are various types of eye surgery. On smaller scales—at the cellular level—light pulses have been used to activate membrane channels and membrane pumps so as to induce changes of electric potential across cell membranes; this has opened prospects of enabling neural circuitry. On even smaller scales—at the sub-cellular level—it is becoming feasible to toggle individual biomolecules from an active to a passive state, and vice versa, by photoswitching of fluorescent labels. The use of femtosecond-duration laser pulses results in high instantaneous peak powers that enable multiphoton (nonlinear) absorption while, concomitantly, avoiding heat damage via single-photon (linear) absorption. Hence, ultrafast laser pulses have also begun to be of utility in a gamut of non-surgical applications: multiphoton laser scanning microscopy with utility in the neurosciences, high harmonic imaging of cells, tissues and organisms, coherent anti-Stokes Raman spectroscopy, and nonlinear imaging of biological entities in general. Intense laser light also has associated with it radiation pressure, and this has been utilized for laser pressure catapulting of dissected samples into appropriate containers for polymerase chain reaction (PCR) investigations. Such laser beams also carry out cell lysis and subsequent catapulting of the contents of the lysed cell into a micropipette for time-resolved capillary electrophoresis, for fusing together of cells and for in-vitro fertilization. Less vigorous applications include the use of ultrafast laser pulses to gently transfect genes into specific cells, with a tightly focused laser beam acting like a sharp but contact-less "needle".

Ultrashort pulses of intense laser light affect matter with which they interact and, in turn, matter also affects the light pulses. Chapter 6 considers both "laser-matter" and "matter-laser" interactions from the biophotonic perspective. Propagation of intense light through matter gives rise to visually spectacular phenomena like supercontinuum generation and filamentation which have helped to probe damage induced in DNA by low-energy interactions involving electrons and OH-radicals. They have also provided high sensitivity diagnostic of stress markers in human saliva. There are openings for similar biophotonics applications without the need for very high-intensity lasers, and possible ways of achieving this are discussed.

Conventional wisdom dictates that quantum mechanical processes do not manifest themselves in large systems, like biological complexes, at room temperature. However, ultrafast time-resolved spectroscopic investigations have begun to yield contradictory evidence: long-range quantum mechanical effects *do* occur in biological systems under physiological conditions. Some topics relevant to the functional role of quantum processes in ultrafast biophotonic phenomena in plants and animals, along with their experimental investigations, are discussed in Chaps. 7 and 8, respectively. The concept of quantum mechanical superposition is briefly presented as are some descriptions of important ultrafast spectroscopy techniques that may be used to explore coherent effects in biology. Chapter 7 considers how coherent energy transfer involving entangled states is much more efficient than classical incoherent energy hopping, and such coherence is utilized by nature in the energy transport mechanism in light-harvesting complexes in plants and in photosynthetic bacteria. One intriguing and counterintuitive manifestation of quantum effects in biology is that of avian navigation in the earth's magnetic field. It appears that such navigation depends on exploiting the earth's magnetic field to tune radical pair production that aids both orientation and navigation. The possibility of complex biological systems performing what is essentially a kind of spin resonance experiment in order to navigate is fascinating; it is discussed in Chap. 8.

The focus in contemporary biology has begun to discernibly shift from morphological explorations and phenotypic probing of organisms to seeking quantitative insights into underlying mechanisms at molecular levels. For instance, protein dynamics are now computationally modelled in terms of trajectories on energy landscapes. These landscapes are multidimensional "cousins" of conventional potential energy surfaces—the variation of potential energy as a function of internuclear coordinates—that physicists and chemists have long used to understand molecular dynamics. As protein trajectories evolve on such landscapes, processes have been discovered that can only be rationalized using quantum mechanical effects like tunneling, and in terms of transitions that occur in non-adiabatic fashion. All these concepts have been creatively adapted by nature so as to be make them useful to the biological milieu. Chapter 9 provides a working overview of such ultrafast processes in proteins like myoglobin.

What opportunities do quantum processes offer for the development of ultrafast artificial systems that succeed in mimicking their natural counterparts? An important question in ultrafast biophotonics, thus, is whether it is possible to incorporate quantum effects into bio-inspired synthetic systems in order to develop devices that possess quantum-enhanced functionality. Chapter 10 explores opportunities of ultrafast photonic functionalities of some proteins in plants and higher organisms, and their potential role in designing new devices that mimic natural systems possessing quantum-enhanced efficiency and adaptability. This chapter discusses the role of ultrafast biophotonics in developing biomimetic devices whose quantum properties can be "engineered" for applications in light-harvesting, solar energy conversion, magnetic field sensing, photonic devices and single-biomolecular electronics.

We end the book by presenting in Chap. 11 an overview of the prospects of forthcoming experimental and theoretical advances in ultrafast biophotonics. The areas discussed include tunneling in biological entities, cavitation in aqueous biological media by thermal as well as plasmonic means, probing biological entities beyond the quantum limit. Emerging techniques that are likely to play a major role in the development of the subject include coherent control of biological processes, femtosecond-long X-ray pulses from free electron lasers for ultrashort imaging of biological materials without causing photodamage, application of terahertz radiation for imaging and spectroscopy, and uses of frequency comb Fourier transform spectroscopy.

Chapter 2
Ultrashort Pulses and Nonlinear Optics: Nuts and Bolts

Abstract In this chapter we summarize some basic concepts related to light, nonlinear optics, and ultrashort pulses that will be employed in our explorations of ultrafast biophotonics. Concepts like the intensity of laser light, ponderomotive energy, nonlinear optics, frequency conversion as well as the characterization and propagation of ultrashort pulses that we discuss here will be frequently encountered in this book.

2.1 Conceptual Aspects

Ultrafast photonics is related to the study of photon-induced phenomena that occur on very short timescales. It is, therefore, but natural that ultrashort pulses of light will form an essential tool for such studies. The basic principle of an ultrashort pulse is quite simple: it is a burst of light energy consisting of only a few cycles of oscillating electromagnetic field [1, 2]. Such pulses are generally characterized by their space- and time-varying electric field $\widetilde{E}(\mathbf{r}, t)$.[1] For the purpose of this book, the quantum mechanical properties of the light field are not important, hence it is sufficient to use the classical approach or Maxwell's equations of electrodynamics [3, 4]. A concise outline of some relevant concepts is provided in the following, with a view to laying the conceptual foundations of what follows in succeeding chapters.

2.1.1 Maxwell's Equations

An appropriate place to start are Maxwell's equations in a homogeneous medium with electric charge density ρ and the electric current density $\mathbf{j}$ which are given by

---

1

- Throughout the book, we use bold letters to represent vector quantities and tilde ($\sim$) to denote quantities rapidly varying in time. Constant or slowly varying quantities are written without the tilde.
- Except where otherwise noted, we use the SI (MKS) system of units.

P. Vasa and D. Mathur, *Ultrafast Biophotonics*, Biological and Medical Physics, Biomedical Engineering, DOI 10.1007/978-3-319-39614-9_2

$$\nabla \cdot \mathbf{E} = \rho, \tag{2.1}$$

$$\nabla \times \mathbf{E} = -\frac{\partial \mathbf{B}}{\partial t}, \tag{2.2}$$

$$\nabla \cdot \mathbf{B} = 0, \tag{2.3}$$

and

$$\nabla \times \mathbf{B} = \frac{\partial \mathbf{D}}{\partial t} + \mathbf{j}. \tag{2.4}$$

Together with the Lorentz force law that relates the force $\mathbf{F}$ acting on a particle with charge q moving with velocity $\mathbf{v}$,

$$\mathbf{F} = q(\mathbf{E} + \mathbf{v} \times \mathbf{B}), \tag{2.5}$$

Maxwell's equations summarize the theoretical content of classical linear electrodynamics and optics. Here, the relation between the $\mathbf{E}$-field and the $\mathbf{D}$-field associated with light is given by

$$\mathbf{E} = \frac{1}{\epsilon_0}(\mathbf{D} - \mathbf{P}), \tag{2.6}$$

with $\mathbf{P}$ being the dipole moment per unit volume or the macroscopic polarization. Similarly, the $\mathbf{B}$-field and the $\mathbf{H}$-field are related by the magnetization $\mathbf{M}$ as

$$\mathbf{B} = \mu_0(\mathbf{H} + \mathbf{M}). \tag{2.7}$$

For topics relevant to this book or for nonmagnetic materials in general, $\mathbf{M} = 0$ is a valid condition, simplifying (2.7) to $\mathbf{B} = \mu_0\mathbf{H}$.

In the case of conventional—or linear—optics, we have

$$\mathbf{P} = \epsilon_0 \chi \mathbf{E}, \tag{2.8}$$

where the complex function, χ, is the linear optical susceptibility. The relation between the macroscopic polarization and the $\mathbf{E}$-field, (2.6) simplifies to $\mathbf{D} = \epsilon_0\epsilon\mathbf{E}$, with the relative dielectric function, $\epsilon = 1 + \chi$.

The real part of the susceptibility describes the physical process of refraction whereas its imaginary part is related to the absorption coefficient, α.

The well-known wave equation for the $\mathbf{E}$-field associated with light can be derived from Maxwell's equations as

$$\nabla^2 \widetilde{\mathbf{E}} - \frac{1}{c_0^2} \frac{\partial^2 \widetilde{\mathbf{E}}}{\partial t^2} = \mu_0 \frac{\partial^2 \widetilde{\mathbf{P}}}{\partial t^2}, \quad (2.9)$$

or, using (2.6), as

$$\nabla^2 \widetilde{\mathbf{E}} - \frac{1}{c^2} \frac{\partial^2 \widetilde{\mathbf{E}}}{\partial t^2} = 0, \quad (2.10)$$

with $c = \frac{c_0}{n}$ being the speed of light in a medium. This is lower than the speed of light in vacuum, $c_0 = \frac{1}{\sqrt{\epsilon_0 \mu_0}} = 2.998 \times 10^8$ m s^{-1}, by a factor n—the refractive index—which, for a nonmagnetic material, is given by

$$n = \sqrt{\epsilon}. \quad (2.11)$$

As with the optical susceptibility or the dielectric function, the refractive index is also a complex function. For topics discussed in this book, we confine our attention to harmonic, linearly polarized, transverse plane wave solutions of (2.10) with angular frequency ω traveling through the medium in the direction along $\mathbf{k}$:

$$\widetilde{\mathbf{E}} = \mathbf{E_0} e^{i(\mathbf{k}\cdot\mathbf{r} - \omega t)} + cc. \quad (2.12)$$

Here, $\mathbf{E_0}$ is the complex **E**-field amplitude and the magnitude of the propagation vector **k** is related to the wavelength of light in vacuum by $k = \frac{2\pi n}{\lambda}$. The corresponding **B**-field amplitude of the light wave is given by

$$\mathbf{B_0} = \frac{1}{c_0} \widehat{\mathbf{k}} \times \mathbf{E_0}. \quad (2.13)$$

Since Maxwell's equations do not involve higher order field terms, (2.10) is a linear, second-order partial differential equation. Hence the superposition principle governing the interference and diffraction of light waves holds. As we see later, interference plays a fundamental role in the generation of ultrashort pulses. Though described here only for linear polarization, the theoretical treatment can be extended to include other polarization states.

Polarization is an important parameter for investigating biological systems. It plays a key role in several fields of biophotonics, like biological nonlinear optics. It is also noteworthy that the range of wavelengths of electromagnetic radiation span a very wide range, from 10^5 m (radio waves) to $\leq 10^{-12}$ m (gamma rays): the wave equation is applicable over the entire range! This scale invariance of classical electromagnetic theory is really remarkable.

2.1.2 Light Intensity and Ponderomotive Energy

One of the most significant properties of electromagnetic waves is that they carry energy and momentum. The light emitted by the nearest stars (other than our sun) travels for millions of kilometres through vacuum to reach the earth, carrying sufficient energy to initiate a chain of photochemical reactions within our eyes. Our eye—and most other light detectors—are not sensitive to rapidly oscillating **E**-fields or **B**-fields but to the cycle-average of the energy density carried by the light waves. The flow of electromagnetic energy per unit time and across unit area is represented by the Poynting vector

$$\widetilde{\mathbf{S}} = \frac{1}{\mu_0}\widetilde{\mathbf{E}} \times \widetilde{\mathbf{B}}. \tag{2.14}$$

Applying (2.14) to plane waves in vacuum, the modulus of the Poynting vector, or the energy density per unit time transported by an electromagnetic wave, is given by

$$\widetilde{S} = |\widetilde{\mathbf{S}}| = \sqrt{\frac{\epsilon_0}{\mu_0}}|\widetilde{\mathbf{E}}|^2, \tag{2.15}$$

which is a time varying quantity. Its cycle-average is the light intensity, I,

$$\langle\widetilde{\mathbf{S}}\rangle = I = \frac{1}{2}\sqrt{\frac{\epsilon_0}{\mu_0}}E_0^2. \tag{2.16}$$

This is also known as the radiant flux density or the optical power density, representing radiant energy incident on—or exiting from—a surface. For a point source, it is proportional to $\frac{1}{r^2}$.

As optical detectors are sensitive to intensity, the phase information about the **E**-field is generally lost in a measurement unless care is taken to employ special techniques based on interference. Like wavelength, the light intensity also spans a wide range of values, as shown in Table 2.1. Ranging from extremely low values ($\sim 10^{-23}$ W cm^{-2}) to extremely high ones ($\sim 10^{30}$ W cm^{-2}), it covers more than fifty orders of magnitude [2]. The lower bound here corresponds to the intensity of the visible part of blackbody radiation at room temperature (300 K), whereas the upper bound corresponds to the Schwinger intensity limit at which the energy content in the light field is sufficient to spontaneously generate an electron-positron pair from vacuum. For comparison, the sun's light intensity on the earth is $\sim$0.1 W cm^{-2} whereas that generated by focussing (to 1 mm^{-2}) light from a reasonably powerful ($\sim$10 mW) continuous-wave laser is $\sim$1 W cm^{-2}. Focusing, pulsed (femtosecond) lasers readily produces intensities in the 10^{15}–10^{16} W cm^{-2} range; at the time of writing this book there are several national facilities across the world which host large laser systems capable of routinely producing intensities of the order of 10^{19}–10^{22} W cm^{-2}.

Table 2.1 Some illustrative order-of-magnitude values of light intensity, the corresponding values of the **E**-field, and the average kinetic energy acquired by an electron due to the ponderomotive potential

Intensity (I) (W m^{-2})	Intensity (I) (W cm^{-2})	Field amplitude (E_0) (V m^{-1})	Cycle average electron kinetic energy at $\lambda = 1.06\ \mu$m (U_p) (eV)	Typical light-matter interaction
10^{34}	10^{30}	10^{18}	10^{17}	Schwinger intensity limit, spontaneous generation of electron-positron pair
10^{24}	10^{20}	10^{13}	10^{7}	Light-induced nuclear fission
10^{22}	10^{18}	10^{12}	10^{5}	Light-induced particle acceleration for cancer therapy
10^{20}	10^{16}	10^{11}	10^{3}	Atomic unit of intensity for an H-atom
10^{18}	10^{14}	10^{10}	13.6	Electron tunneling from an atom
10^{10}	10^{6}	10^{6}	10^{-7}	Typical intensities at which nonlinear effects are observed
10^{6}	10^{2}	10^{4}	10^{-11}	A continuous wave laser intensity that can cause serious burn injuries
10^{3}	10^{-1}	10^{3}	10^{-14}	Total intensity of the sun on the earth's surface
10^{2}	10^{-2}	10^{2}	10^{-15}	Thermal radiation from a human body
10^{-6}	10^{-10}	10^{-2}	$<10^{-15}$	Total intensity of the cosmic background radiation at 2.8 K
10^{-11}	10^{-15}	10^{-4}	$<10^{-15}$	Vision threshold of the human eye
10^{-18}	10^{-22}	10^{-8}	$<10^{-15}$	Intensity of the visible part of blackbody radiation at 300 K

Light is absorbed and emitted in discrete numbers of photons, but even from a very low intensity source, like our Sun, light reaching us on earth has such a large photon flux ($\sim 10^{10}$ photons s^{-1} m^{-2}) that its inherently discrete nature is totally obscured: what is generally observed—and considered by the classical theory of electrodynamics—is a continuous phenomenon. An alternate viewpoint to estimate the strength of the light field is to consider the amplitude of the associated **E**-field. Here, a useful benchmark is that of the Coulomb field, E_{at} within a hydrogen atom whose Bohr radius a_0 is 0.53×10^{-10} m when the electron is in its ground state:

$$E_{at} = \frac{e}{4\pi\varepsilon_0 a_0^2} \sim 5 \times 10^{11}\ \text{V m}^{-1}. \tag{2.17}$$

The Coulombic field experienced by this electron maps to an intensity of $\sim 10^{16}$ W cm^{-2}. At this intensity value, as depicted in Fig. 2.1, a non-resonant light field is sufficiently strong to ionize a hydrogen atom. Strong optical fields can distort (or "dress")

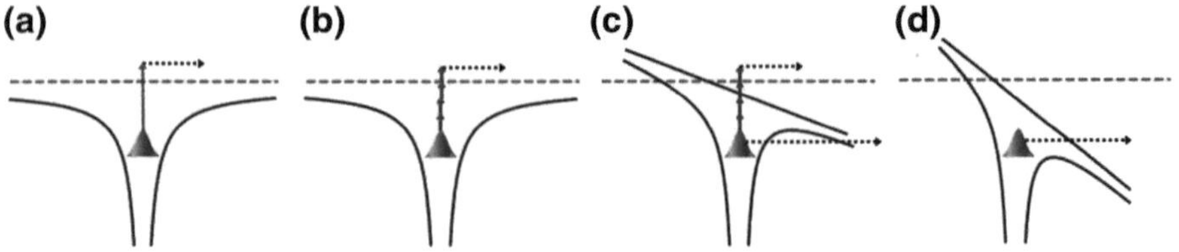

Fig. 2.1 **a** Direct (one-photon resonant), **b** multiphoton, **c** tunnel and **d** over-the-barrier ionization of an electron. At sufficiently high intensity, field dressing of the atomic potential, as schematically depicted in (**c**) and (**d**), becomes the dominant effect

the Coulombic field such that tunneling through the Coulomb barrier can occur. Indeed, ionization of matter is inevitable under strong-field (high laser intensity) conditions.

The electron that tunnels through the dressed potential (Fig. 2.1) carries signatures of the extent to which the strong laser light has distorted the Coulombic potential and, hence, electron spectroscopy proves to be an important tool in gaining insights into strong field ionization. The theoretical underpinnings of strong field ionization were laid by Volkov as long ago as 1935, decades before the advent of strong optical fields. Volkov solved the Dirac equation in the presence of a plane wave [5] and, along with subsequent pioneering theoretical considerations of electron motion in intense, time-varying electromagnetic fields [6–11], helped lay the foundations on which modern strong field science has developed. We shall have recourse to using some of these strong field concepts and techniques in discussions pertaining to ultrafast biophotonics in later chapters. Today, pulse durations and laser intensities are readily available that can easily generate fields much larger than E_{at}. Whenever the **E**-field associated with light is comparable to E_{at}, the material response does not follow (2.8); in fact, it becomes dependent on higher powers of the **E**-field. This is the domain of nonlinear optics and strong field science [12]. We shall discuss several novel and unusual effects that arise due to nonlinear optical responses and explore their applications in ultrafast biophotonics. During the course of this book, we shall mostly encounter intensity values in the range 10^6–10^{15} W cm^{-2}. We shall also see that ultrashort pulses are essential to reach these intensity values.

As any light-matter interaction involves motion of electrons, it is useful for our study of nonlinear effects to understand electron dynamics in a strong field. Unless the electron is moving at a very high speed, approaching c_0, (2.5) and (2.13) confirm that the influence of the magnetic field associated with light is negligible. For a linearly polarized **E**-field given by (2.12), the displacement **x** of a free electron of mass m_e can be written as

$$m_e \ddot{\mathbf{x}} = -e\widetilde{\mathbf{E}}, \tag{2.18}$$

giving the cycle-averaged kinetic energy of an electron as

$$U_p = \frac{e^2 E_0^2}{4 m_e \omega^2}. \tag{2.19}$$

This energy, which is associated with the oscillatory motion of the electron induced by the optical field is known as the ponderomotive energy. It is a strongly frequency dependent quantity and takes the value of 13.6 eV for $E_0 \sim 0.1E_{at}$ at $\lambda = 1.06$ μm. The ponderomotive energy corresponding to some typical threshold values of the light intensity are listed in Table 2.1.

The standard model of strong field ionization, the Keldysh-Faisal-Reiss (KFR) theory [13–15], considers electron ejection from an atom in a strong non-resonant ($\hbar\omega \ll I_e$) optical field in terms of either multiphoton (MPI) or tunnelling ionization (TI), schematically shown in Fig. 2.1. The boundary between the two regimes is established by considering the adiabaticity parameter (the Keldysh parameter), $\gamma_k = \sqrt{\frac{I_e}{2U_p}}$. γ_k is simply the ratio of the time taken by an electron to tunnel through the field-distorted Coulomb barrier to the half-optical period of the field causing the distortion. In the expression for γ_k, I_e is the ionization energy of the atom. Experimentally, γ_k values are readily controlled by (i) varying the laser intensity or by altering the frequency of the incident laser light (in other words, altering the value of U_p) and (ii) by changing the target atom (in other words, "replacing" the value of I_e). Sufficient experimental evidence has now been accumulated to establish that TI dominates the ionization dynamics when $\gamma_k \ll 1$, wherein there is a significant distortion of the atomic potential by the optical field and the energy distribution of the ejected electron is largely determined by its classical propagation in the oscillating field. The MPI picture appears to be more appropriate in describing the ionization dynamics when $\gamma_k > 1$. In this regime, the electron energy spectrum contains signatures of atomic resonances and so-called above-threshold ionization (ATI) peaks. For complex structures like proteins and DNA, and for negative ions, the ionization energy can be small enough to ensure that even an electric field very much smaller than E_{at} is sufficient to give rise to TI at infrared (IR) frequencies.

As already noted, strong fields are generated using intense laser pulses: at the leading edge of such a pulse the intensity experienced by the atom or molecule is likely to be such that MPI dominates the ionization dynamics whereas closer to the peak of the pulse it is TI that drives most of the ionization. If the pulse is of long enough duration, one may even envisage a situation in which a series of MPI events at the rising edge of the pulse ionize all the electrons in the atom or molecule and no further ionization is possible. The discerning reader might have also noted another possibility: in a multielectron atom or molecule, the rising edge would eject an electron by MPI so that subsequent ionization is not of the neutral atom or molecule but of the corresponding ion which, as the intensity within the pulse rises, is subjected to further ionization. Such sequential ionization is a ready consequence of both MPI and TI but one that may be possible to "switch off" by opting for ultrashort laser pulses of sufficient intensity.

2.1.3 Emission of Light and Dispersion

Accelerating charges emit electromagnetic radiation [3, 4]. An oscillating dipole, a charged particle, like an electron traveling along a curved path—as in a synchrotron or an electron in an atomic orbital—are examples of accelerating charges. All of them emit electromagnetic radiation. The simplest light emitting mechanism to visualize is the oscillating dipole—two opposite charges vibrating along a straight line. The most significant mechanism for emission and absorption of light is the rearrangement of weakly bound electrons in atoms and molecules. The emission and absorption of light, although a quantum mechanical process, can be adequately explained in terms of the classical oscillating dipole. A material medium is envisioned as an ensemble of a large number of closely spaced atomic dipoles, with the spacing being much smaller than the wavelength of light. When a linearly polarized light wave impinges on such a medium, a bound electron in an atom can be represented as a classical forced oscillator that is driven by the time-varying **E**-field [3, 4, 12, 16]. The electrostatic force binding the electron provides the restoring force ($\mathbf{F_r} = -k_r\mathbf{x} = -m_e\omega_0^2\mathbf{x}$) which opposes the periodic driving force ($-e\widetilde{\mathbf{E}}$). As in (2.18), the equation of motion is given by

$$m_e\ddot{\mathbf{x}} = -2m_e\gamma\dot{\mathbf{x}} - e\widetilde{\mathbf{E}} - m_e\omega_0^2\mathbf{x}, \tag{2.20}$$

where ω_0 is the resonance frequency of the atom or molecule. Equation 2.20, known as Lorentz model of the atom, represents a highly idealized and simplified classical model for the most fundamental atomic dipole light-matter interaction. Even under illumination of bright sunlight, the amplitude of oscillation is $\sim 10^{-17}$ m, which is much smaller than a_0. For more intense excitation, $x \geq a_0$ and restoring force terms dependent on higher powers of displacement have to be considered. A damping term $2m_e\gamma\dot{\mathbf{x}}$ is added phenomenologically to avoid an unphysical resonance condition or the infinitely large amplitude (spontaneous ionization) when $\omega \sim \omega_0$. Complete understanding of the damping term or the relaxation processes is possible only through the theory of quantum electrodynamics and is beyond the scope of this book.

The total polarization induced in the medium due to the incident light is given by

$$\widetilde{\mathbf{P}} = -Ne\widetilde{\mathbf{x}} = \frac{\frac{Ne^2}{m_e}}{\omega_0^2 - i\omega\gamma - \omega^2}\widetilde{\mathbf{E}}, \tag{2.21}$$

where N is the number density of atoms in the medium. The complex coefficient between $\widetilde{\mathbf{P}}$ and $\widetilde{\mathbf{E}}$ is defined as the optical susceptibility $\chi(\omega) = \frac{\omega_{pl}^2}{\omega_0^2 - i\omega\gamma - \omega^2}$ that we introduced earlier. Here, $\omega_{pl}^2 = \frac{Ne^2}{m_e\epsilon_0}$ is the characteristic plasma frequency. The complex function representing the frequency dependence of $\chi(\omega)$ is a Lorentzian whose imaginary part is related to the absorption coefficient, $\alpha(\omega) = \frac{\omega}{n_0 c_0} Im[\chi(\omega)]$, where n_0 is the background refractive index. Like the optical susceptibility and the

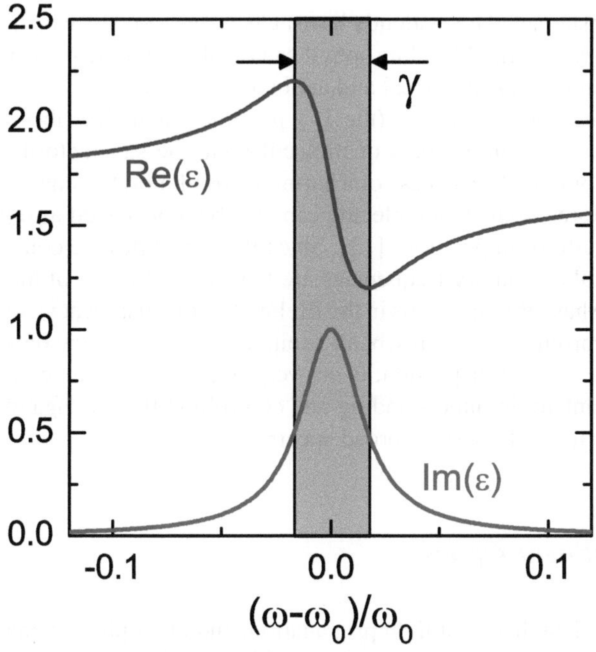

Fig. 2.2 Dispersion of the real and imaginary parts of the dielectric constant near a resonance. The *highlighted area* marks the region of anomalous dispersion

dielectric constant, the refractive index is also a complex quantity. The dispersion of the real and imaginary part of the dielectric function is shown in Fig. 2.2. Its real part governs the refraction of light at the interface of two media and is given by $\sqrt{1+Re[\chi(\omega)]}=\sqrt{Re[\epsilon(\omega)]}$. As illustrated in Fig. 2.2, far away from the material's resonance frequency, the imaginary part of $\epsilon(\omega)$ or $n(\omega)$ is negligible. Hence, in this regime the frequency dependent refractive index, or the dispersion of n, can be expressed as

$$n^2(\omega) = 1 + \frac{\omega_{pl}^2}{\omega_0^2 - \omega^2}. \tag{2.22}$$

This spectral range is known as the normal dispersion, or transparent regime, characterized by $\frac{\partial n}{\partial \omega} > 0$. Close to resonance, the imaginary contribution becomes significant, yielding enhanced absorption. Here, $\frac{\partial n}{\partial \omega} < 0$, and is known as anomalous dispersion. Though highly simplified, the Lorentz atomic model can adequately explain the optical response of a wide range of condensed media—such as liquids, dielectrics, metals, semiconductors, birefringent crystals, and diverse biological entities. Most dielectric or transparent materials, like glass and silica (SiO_2), are characterized by anomalous dispersion in the ultraviolet (UV) and/or IR wavelength ranges, whereas chlorophyll or stained glass evidently have at least one resonance in the visible part of the spectrum. A material like chlorophyll can exhibit multiple resonances, implying several spectral regions of normal and anomalous dispersion.

The Lennard-Jones potential (the L-J potential) that describes the interaction between a pair of neutral atoms or molecules can be satisfactorily approximated as a parabolic potential, as in case of a harmonic oscillator, for small interatomic distances. Therefore, atoms in a molecule can also be represented as dipoles vibrating about their equilibrium positions [12]. Since the molecules are comparatively massive, the natural oscillatory frequencies are low, in the IR part of the spectrum. For example, H_2O has vibronic bands in the IR, but it is transparent at visible frequencies.

For monochromatic or narrowband excitation, the frequency dependence of the refractive index, or the dispersion, is not very important. However, as we shall see, it is a key parameter in understanding and controlling the propagation of ultrashort pulses that intrinsically possess broad spectra.

2.1.4 Nonlinear Optics

The usual classical theory of light propagation, and other phenomena like reflection and refraction that are derived on the basis of Maxwell's equations, assumes a linear relationship between the macroscopic polarization and the electric field strength [12], as given in (2.8) and (2.21). In other words, the force acting on the atomic oscillators in the presence of an optical field is considered to be linearly dependent on the displacement from the mean position (2.20). However, just as a mechanical harmonic oscillator can be overdriven into a nonlinear response by applying sufficiently large force, atomic or molecular oscillators can also be overdriven by intense optical fields.

Nonlinear optics is the study of such light-matter interactions for which higher powers of electric field play a dominant role. It is the study of phenomena that occur as a consequence of the modification of the optical properties of a medium by the presence of light (usually intense light). Generally, only laser light and ultrashort pulses are sufficiently intense to overdrive the atomic oscillators. In the linear optical regime, the optical response—or the susceptibility—is a material specific quantity that is independent of the applied field strength. Here, $\widetilde{\mathbf{P}}$ increases linearly with $\widetilde{\mathbf{E}}$. In the case of intense fields, this linear relationship cannot be maintained. The polarization generated in a material medium is then described by expressing (2.8) as a power series in field strength $\widetilde{\mathbf{E}}$:

$$\begin{aligned}\widetilde{P} &= \epsilon_0(\chi^{(1)}\widetilde{E} + \chi^{(2)}\widetilde{E}^2 + \chi^{(3)}\widetilde{E}^3 + \cdots), \\ &= (\widetilde{P}^{(1)} + \widetilde{P}^{(2)} + \widetilde{P}^{(3)} + \cdots.\end{aligned} \tag{2.23}$$

The quantities $\chi^{(1)}$, $\chi^{(2)}$ and $\chi^{(3)}$ represent, respectively, the linear (first-), second- and third-order optical susceptibility whereas $\widetilde{P}^{(i)}$, $i = 1$ to 3, represents the corresponding polarization [12, 17]. Each higher-order term of the series expansion gives rise to distinct physical processes. We will briefly describe a few of these nonlinear effects that will be of relevance to ultrafast biophotonics. For simplicity we shall consider the electric field and polarization to be scalar quantities (the scalar approximation). We shall also assume that the polarization depends only on the instantaneous electric field strength and that the medium responds instantaneously. In presence of dispersion and losses, (2.23) needs to be generalized. Such a comprehensive treatment of the nonlinear polarization is beyond the scope of this book and can be found in standard textbooks on nonlinear optics [12, 17].

The typical value of linear susceptibility for common dielectrics is close to unity, whereas the values of $\chi^{(2)}$ and $\chi^{(3)}$ that would generate polarization equal to the linear term for the extreme field strength of E_{at} are $\sim 1.9 \times 10^{-12}$ m V^{-1} and $\sim 3.8 \times 10^{-24}$ m^2 V^{-2}, respectively [12, 17]. Since these values are so much smaller than χ, the higher order polarization terms do not contribute significantly unless the electric field is sufficiently high. Under very intense excitation where, as already noted, ionization effects dominate the light-matter interaction, the series expansion in (2.23) may not converge. To mathematically describe propagation through a nonlinear medium, the wave equation (2.9) has to be modified to include the higher-order time varying polarization terms. These terms act as sources of additional components of electromagnetic fields resulting from the nonlinear interaction and play a key role in ultrashort pulse propagation through media. A complete analysis of pulse propagation through nonlinear media is very demanding, owing to the large number of effects involved and, as seen in Chaps. 5, 6, extensive use has to be made of approximations and phenomenological models. In the following sections we provide a brief qualitative description of a number of nonlinear optical processes that we shall have recourse to refer to in later chapters.

2.1.5 Second-Order Nonlinear Effects

We begin by considering an optical field comprising two distinct frequency components being incident on a second-order nonlinear optical medium [12, 17]. The incident field and the second-order polarization generated within the medium are, respectively, given by

$$\widetilde{E} = E_1 e^{-i\omega_1 t} + E_2 e^{-i\omega_2 t} + cc, \tag{2.24}$$

$$\begin{aligned}\widetilde{P}^2 = \epsilon_0\chi^{(2)}(E_1^2 e^{-i2\omega_1 t} + E_2^2 e^{-i2\omega_2 t} + cc)\ (\text{SHG}) +\\ 2\epsilon_0\chi^{(2)}(E_1 E_2 e^{-i(\omega_1+\omega_2)t} + cc)\ (\text{SFG}) +\\ 2\epsilon_0\chi^{(2)}(E_1 E_2^* e^{-i(\omega_1-\omega_2)t} + cc)\ (\text{DFG}) +\\ 2\epsilon_0\chi^{(2)}(E_1 E_1^* + E_2 E_2^*)\ (\text{OR}).\end{aligned} \tag{2.25}$$

The second-order nonlinear polarization consists of five distinct frequencies [18–20]. Two of them correspond to the second harmonic of the incident frequency, known as second harmonic generation (SHG). The other three include generation of sum frequency (SFG), difference frequency (DFG) and optical rectification (OR), in which a DC or static field is generated within the medium. In physically realistic situations, however, only one of these components is generated with appreciable intensity. The reason for this frequency selectivity is that the nonlinear polarization can efficiently radiate at a particular frequency only if energy as well as momentum conservation conditions are satisfied. The energy is conserved by destruction of incident photons and creation of photons at new frequencies, whereas momentum is conserved if there is matching of the refractive indices at the incident and created frequencies. The latter requires a phase-matching condition to be satisfied, which is usually possible only for one of the frequency components of the nonlinear polarization. The polarization of the incident light and the orientation of the nonlinear medium (for example, a crystal like β-BaB_2O_4 or KDP) are important in selecting the frequency component. The efficiency of the conversion of incident frequency into converted frequency (SHG/SFG/DFG) depends strongly on experimental conditions and on the nonlinear medium. Under proper experimental conditions, the process of SHG can be so efficient that nearly all of the power in the incident beam at frequency ω is converted to radiation at frequency 2ω.

Out of the four processes, SHG is the most widely used in biophotonics [21, 22], including microscopy (Chap. 3). Its common use is to convert the output of an IR laser—such as a Nd:YAG laser (1064 nm) or a Ti:sapphire laser ($\sim$700–900 nm)—to the green or blue regions of the spectrum. SHG can be thought of as annihilation of two photons of frequency ω and simultaneous generation of a photon of frequency 2ω in a single quantum-mechanical process. Similarly, SFG involves annihilation of photons at ω_1 and ω_2 accompanied by creation of photon at $\omega_1 + \omega_2$. In DFG, the higher frequency pump-photon (ω_1) is converted into the difference frequency idler ($\omega_1 - \omega_2$) in presence of the signal frequency (ω_2). This results in generation of additional radiation at ω_2, amplifying the already present signal component. In other words, the conversion to idler frequency radiation is stimulated by the amplification of the signal frequency. Hence, DFG is also known as optical parametric amplification and we shall make use of this process in studies on DNA damage (Chap. 6).

SFG and optical parametric amplification are regularly used to produce tunable UV and IR radiation, respectively, by mixing the output of a tunable visible laser with that of a fixed frequency near-IR laser. Frequency tuning is achieved by adjusting the phase-matching condition. Multiple stages of SHG, SFG and parametric amplification can be combined within a single source to realize a tunable source that spans a wide spectral range from UV to IR. Second-order nonlinear processes can also be used to generate pulsed terahertz (THz) radiation that is likely to potentially possess enormous utility in biophotonics applications in the coming years. THz radiation is also generated in the course of filamentation, a phenomenon that accompanies the propagation of high-intensity ultrashort laser pulses through matter (Chap. 6).

All four nonlinear processes described so far are examples of parametric, or lossless, processes. These are quantum mechanical processes in which the photon energy is conserved [12, 17] and are represented by real nonlinear susceptibility. There can also be nonparametric nonlinear processes in which the photon energy is not conserved. Such distinction also exists in the linear optical regime. Processes like reflection and refraction that are governed by the real component of susceptibility are examples of parametric processes, whereas absorption—represented by the imaginary component—is an example of a nonparametric process. In the latter, the energy is converted into material excitation or heat and the photon energy is not conserved. The other difference between the two types of processes is that the parametric processes are instantaneous, having response time governed by the Fourier transform of the optical response in the transparent spectral range. The nonparametric process involves population transfer from the ground state of the medium to an excited state, which then takes a finite amount of time (ranging from several femtoseconds to even milliseconds) to relax back.

The generation of other frequencies in a nonlinear interaction can also be thought of as a consequence of an overdriven atomic or molecular oscillator. For small changes around the equilibrium position, both the atomic potential as well as the L-J potential can be approximated as a parabolic potential (Fig. 2.3). The motion of an electron in such a potential is described by a sinusoidal function at the excitation frequency. In presence of intense optical excitation, the amplitude of oscillation becomes larger and the potential experienced by the electron deviates from being parabolic. The electron motion in such a deformed potential continues to remain periodic but becomes distinctly non-sinusoidal. The dynamics can then be described by expanding the response in a Fourier series with non-zero components at harmonic frequencies. Since the accelerating charge emits electromagnetic radiation, the components of electron motion at higher harmonic frequencies act as the source for higher order polarization that emit frequencies other than the excitation frequency [12].

The shape of the potential is strongly governed by the atomic (or molecular) structure. In condensed media possessing inversion symmetry, the potential is an even function of the inter-atomic distance and, hence, only odd-order nonlinear effects will be observed. Therefore, liquids and materials like glass do not exhibit second-order effects except from the surface, where the symmetry is broken.

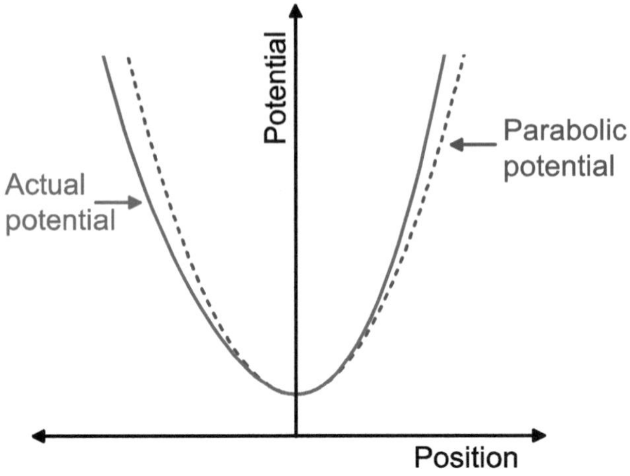

Fig. 2.3 Potential energy curve for a non-centrosymmetric medium. Only for weak optical excitation does the electron experience a parabolic potential

2.1.6 Third-Order Nonlinear Effects

Just as the second-order susceptibility is seen to mix three different fields, the third-order susceptibility can mix four different electric fields [12, 17, 23]. Therefore, in the most general case, the expression for $\widetilde{P}^{(3)}$ is very complicated, with several different components. For a simplified case, where all three fields are monochromatic and identical, $\widetilde{E} = E_0 e^{-i\omega t} + cc$, the polarization is given as

$$\widetilde{P}^3 = \frac{1}{8}\epsilon_0 \chi^{(3)}(E_0^3 e^{-i3\omega t} + cc)\ (\text{THG}) + \frac{3}{8}\epsilon_0 \chi^{(3)}(E_0^3 e^{-i\omega t} + cc)\ (\text{Intensity dependent response}). \tag{2.26}$$

The first term in (2.26) describes third harmonic generation (THG), in which three photons at frequency ω are destroyed and one photon of frequency 3ω is created [12, 24, 25]. THG can be used to generate UV frequencies from IR lasers. The second term describes the component at the incident frequency, leading to an intensity dependent optical response. It can be described as an intensity dependent refractive index experienced by the incident wave, represented as

$$n(I) = n + n_2 I, \tag{2.27}$$

where $n_2 = \frac{3}{2n^2\epsilon_0 c_0}\chi^{(3)}$ characterizes the strength of the optical nonlinearity. The intensity dependent refractive index gives rise to several phenomena like self-focusing, self-phase modulation, saturation of absorption, and two-photon absorption [12, 17]. We shall frequently encounter such nonlinear refractive indices in later discussions of biophotonic phenomena involving diverse biological entities, such as water, DNA, and proteins in human saliva (Chap. 6). Not all of these processes correspond to a real refractive index. Accordingly, the expression (2.27) has to be generalized to include the real as well as the imaginary optical response, in other words the parametric and nonparametric optical processes. Apart from the electronic nonlinearity arising due to non-parabolic potential—known as Kerr nonlinearity—other mechanisms like plasma generation, filamentation, Brillouin scattering and Raman scattering processes can also induce third-order nonlinear responses [12, 17, 23].

One of the most frequently encountered manifestations of $\chi^{(3)}$ is self-focusing [26, 27]. It occurs when a beam of light with non-uniform transverse intensity profile propagates through a nonlinear medium with $n_2 > 0$. The non-uniform intensity profile of the incident beam maps to a non-uniform refractive index profile such that the medium starts to act like a convex lens, causing the beam to converge. As the beam converges, its intensity increases and, concomitantly, so does the extent of self-focusing. An intensity catastrophe is averted when the intensity reaches a high enough value to induce ionization of the medium. The free electrons that are thus generated contribute a negative sign to the overall value of n_2 and, consequently, defocusing occurs. Propagation of high intensity optical pulses thus progresses in a series of self-focusing—plasma defocusing—self-focusing cycles [28]. The entire process gives rise to formation of filaments within the medium and we shall have recourse to involving such propagation dynamics in later chapters in connection with laser eye surgery (Chap. 5) and filamentation-induced damage to DNA (Chap. 6) in physiologically relevant aqueous media [29, 30].

It is of interest to note that at the focal point, the intensity of the self-focused pulse can be sufficiently high to cause optical damage; this has been put to use by creating microfluidic and nanofluidic channels within materials [31] and for cell alignment studies [32].

The propagation through a nonlinear medium also results in an intensity dependent phase [12, 17, 23] given by $\phi(z, t) = n_2 kzI(t)$. This effect is known as self-phase modulation (SPM). Since the rate of change of the optical phase corresponds to frequency, SPM leads to generation of new frequency components governed by $\frac{dI(t)}{dt}$. For ultrashort pulses, the intensity varies on femtosecond timescales. Therefore, SPM leads to a significant widening of the spectrum as the pulse propagates through the medium. Apart from SPM, a collection of other third-order nonlinear processes like self-steepening, cross-phase modulation and stimulated Raman scattering also induce substantial spectral broadening upon intense pulse propagation. The visually prominent manifestation of these phenomena is known as supercontinuum (SC) generation [23, 33–35].

For investigations of SC generation in transparent liquid media, water is a natural choice due to its importance in biological systems [36, 37]. A typical SC spectrum generated in water by intense, IR ultrashort pulses is shown in Fig. 2.4. At relatively lower fluence, the spectral broadening is caused by non-parabolic potential or Kerr nonlinearity corresponding to $n_2 \sim 4 \times 10^{-16}$ cm^2 W^{-1}. Increasing the fluence

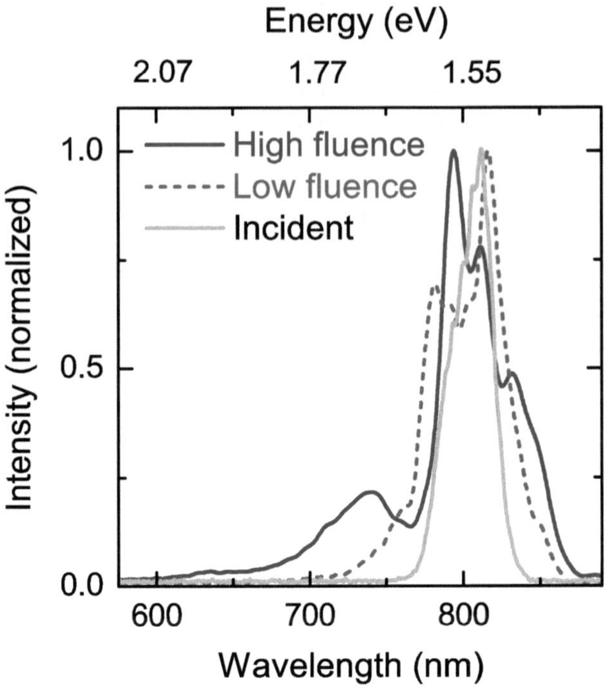

Fig. 2.4 Supercontinuum (SC) spectra generated in water by intense, infrared ultrashort pulses at low and high fluence. At higher fluence, there is an additional broadening, in the visible range due to plasma generation (Chap. 6)

results in plasma generation and filamentation leading to additional contribution to the nonlinearity and, hence, substantial increase in the SC spectral width (Chap. 6). Under optimized conditions, in specially designed photonic crystal fibres, the SC spectrum can span an extremely wide spectral range—over three octaves—and this has played an important role in the development of broadly tunable ultrafast light sources [38, 39], generation of few-cycle pulses [40], as well as the Fourier synthesis of unconventional optical waveforms [41, 42]. SC sources are particularly of interest for ultrafast biophotonics because of their potential applications in imaging and microscopy [43, 44].

Intensity dependent nonparametric processes result in either enhanced absorption or saturation of absorption. Typically, a medium has a finite number of atoms or molecules that can absorb light. Under intense illumination, a large fraction of these atoms/molecules absorb light and are transferred to an excited state, depleting the number available for further absorption of light. This effect is known as saturation of absorption [12, 17]. Intense illumination can also open up new absorption channels, resulting in induced absorption. One of the examples of the latter effect is the process of two-photon absorption [21, 22]. In the linear regime, the atom/molecule is transparent to the excitation frequency. However, under sufficiently strong illumination, it can make a transition from its ground state to an excited state by simultaneous absorption of two laser photons. Once in the excited state, it can relax via emitting a photon whose frequency is higher than the excitation frequency. This process is known as two-photon fluorescence (or two-photon photoluminescence) and it plays a crucial role in the development of high resolution optical microscopy for biological applications.

Since third-order nonlinear interactions generate an intensity dependent optical response, their manifestation is very useful for performing time-resolved studies with ultrashort pulses. Several experimental techniques have been developed to investigate the temporal evolution of the system following an impulsive optical excitation, and will be frequently encountered in this book.

2.1.7 Higher-Order Nonlinear Effects

Under intense optical excitation—approaching E_{at}—we have already noted the breakdown of the perturbative expansion of the nonlinear polarization given in (2.23). The field is strong enough to cause multiphoton nonlinear effects and to distort—or dress—the atomic potential. As discussed in Sect. 2.1.2, different ionization processes dominate this regime. This strong field also accelerates the ionized electron because of which new optical frequencies can be generated. High-harmonic generation (HHG) is an example of such a higher-order nonlinear process in which all odd harmonics of the laser frequency, up to some cut-off (can be a few hundred harmonics), are emitted by an atomic medium [45, 46]. HHG occurs due to the oscillatory motion of the ionized electron in the field of the deformed atomic or molecular ions. One of the principal applications of HHG is to produce radiation at much higher frequencies, corresponding to the extreme UV (XUV) and X-ray spectral regions, for the generation of attosecond pulses [47].

As is discussed in Chap. 6, there is a considerable contemporary interest in the physics of higher-order nonlinear effects and their applications to biological systems. The drive to investigate intense field effects in the context of biophotonics is fuelled by a host of potential applications in, among others, cancer therapy and DNA modification.

2.1.8 Ultrashort Pulses

Until about the beginning of this century, any laser pulse that lasted for 100 fs or less was deemed to be ultrafast. Definitions have begun to change and, now, an ultrashort pulse is one which comprises only a few cycles of an electromagnetic wave. Such pulses are generated by a coherent superposition of a large number of plane waves and are usually described as a product of an oscillatory component (the actual time variation of the optical field) and a pulse-envelope function, generally taken as Gaussian [1, 2]. Therefore ultrashort pulses are not very different from the plane waves considered so far in this chapter and the topics considered till now continue to be relevant. In spite of this similarity, few-cycle pulses offer unique advantages in terms of achieving, simultaneously, high temporal and spatial resolution, generation of intense optical fields, and large spectral bandwidth. These ultrafast pulses make possible investigations of light-matter interactions on timescales in which the nuclear dynamics in molecules are effectively "frozen". From an applications viewpoint, they open the possibility of high-speed data transmission and processing of information. As we shall see in subsequent chapters, new issues will be encountered that relate to the generation, characterization, and propagation of such pulses. In this book, we shall mainly focus on ultrafast optics in the visible and lower-frequency spectral bands and on time scales down to a few femtoseconds although future prospects of probing attosecond dynamics in biophotonics are briefly discussed in Chap. 11.

The electric field of a light pulse is a function of both space and time but, in most cases that are of relevance to ultrafast biophotonics, we shall be mainly interested in its temporal characteristics. For simplicity, we shall treat the electric field as linearly polarized, and represent only by its magnitude (the scalar approximation). Hence, the electric field of a pulse whose central frequency is ω_0 is expressed as

$$E(t) = \frac{1}{2}\sqrt{I(t)}e^{-i[\omega_0 t+\phi(t)]} + cc, \tag{2.28}$$

where $\sqrt{I(t)}$, is the time varying amplitude envelope and $\phi(t)$ is the temporal phase. Generally, $I(t)$ and $\phi(t)$ vary slowly compared to the oscillating part except for few-cycle pulses [1, 2]. To further simplify the mathematical expression, the rapidly oscillating $e^{-i\omega_0 t}$ term and the complex conjugate are often neglected under the analytic signal approximation, giving

$$E(t) = \sqrt{I(t)}e^{-i\phi(t)}. \tag{2.29}$$

For a Gaussian-intensity pulse, the time varying amplitude is expressed as $E_0 e^{[-2ln2(\frac{t}{\tau_{FWHM}})^2]}$, where τ_{FWHM} represents the pulse width (full width at half maximum) and E_0 is the maximum field amplitude. The expression of the complex electric field in the frequency domain, $\mathcal{E}(\omega)$, is obtained by Fourier transforming $E(t)$ as

$$\mathcal{E}(\omega) = \sqrt{S(\omega)}e^{-i\varphi(\omega)}. \tag{2.30}$$

Here, $S(\omega)$ represents the spectrum and $\varphi(\omega)$ is the spectral phase. The electric field, intensity and (real) amplitude of a Gaussian-intensity pulse (τ_{FWHM} of 10 fs) are depicted in Fig. 2.5a. The corresponding spectrum is shown in Fig. 2.5b.

Since the temporal and spectral domains are related via a Fourier transform, the widths in the two domains are also related via the uncertainty principle: a narrower pulse in the temporal domain corresponds to a wider spectrum, and vice versa.

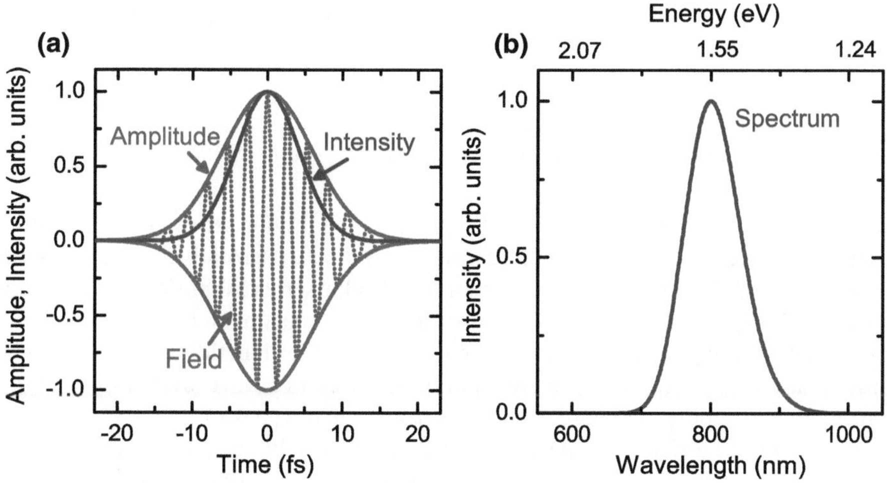

Fig. 2.5 **a** The temporal electric field, its amplitude (real-part) and intensity of a transform limited, near-IR, ultrashort Gaussian pulse with FWHM= 10 fs where FWHM corresponds to the intensity width at half maximum. **b** The corresponding spectrum centred at 1.55 eV (800 nm). The phase of this pulse in the time or frequency domain is a constant and is not plotted

The product of the temporal and spectral widths of pulse is referred to as the time-bandwidth product (TBP), a dimensionless quantity representing the figure of merit of an ultrashort pulse [1]. The smaller the TBP, the simpler is the pulse structure. The TBP for a Gaussian pulse is 0.441. This value represents the shortest pulse that can be generated for a given spectral width. For any pulse shape, the minimum TBP corresponds to the condition when the temporal and spectra phase are constant, that is, there is no change in the frequency within the pulse duration. Such a transform limited Gaussian pules is depicted in Fig. 2.5a. As we will see in the following section, material dispersion leads to pulse broadening, and, consequently, an increase in the TBP. In ultrafast biophotonics, efforts are always aimed at maintaining the lowest possible TBP.

2.1.9 *Propagation of Ultrashort Pulses*

A large number of co-propagating spectral components is indispensable for the generation of an ultrashort pulse. We describe the propagation of such a wavepacket in terms of group velocity, $v_g(\omega) = \frac{c_0}{n_\omega}\left[1 + \frac{\omega}{n_\omega}\frac{dn}{d\omega}\right]^{-1}$. When this wavepacket propagates through a medium, each frequency component acquires a different phase due to the material dispersion or the frequency dependent refractive index (n_ω).Under such conditions, it is customary to represent the spectral phase in a Taylor series expansion, such as

$$\varphi(\omega) = \varphi_0 + (\omega - \omega_0)\varphi_1 + \frac{1}{2!}(\omega - \omega_0)^2\varphi_2 + \frac{1}{3!}(\omega - \omega_0)^3\varphi_3 + \cdots, \quad (2.31)$$

where φ_0 is a constant and φ_i, $i > 0$ represents the Taylor series coefficient proportional to the ith derivative (with respect to ω) of n_ω. Alternatively, the temporal phase $\phi(t)$ can also be represented in a similar Taylor series expansion [1, 2]. In the absence of dispersion and other phase distortions, the spectral and temporal phase is a constant, yielding a transform limited pulse. The first term in the series, the zeroth-order phase (φ_0 or ϕ_0), is the relative phase of the carrier wave with respect to the envelope. It is simply the phase of the carrier wave at the peak of the pulse envelope and is known as carrier envelope phase (CEP). As shown in Fig. 2.6a, when the pulse is several-cycles long, the variation in CEP from 0 to π only slightly shifts the carrier wave away from the peak of the envelope. The real part of the electric field of a 10 fs (τ_{FWHM}) pulse with CEP values of 0 and π look almost identical. However, when the pulse is only one- or two-cycle long, the CEP does matter (Fig. 2.6b)! All three values of CEP shown in Fig. 2.6b correspond to the same amplitude profile but have very different field values at the peak of the field. There are several nonlinear processes, like TI (Fig. 2.1), which depend on the sign of the field and hence exhibit strong CEP dependence when experiments, for example, on atomic or molecular ionization, are performed with few-cycle pulses [48]. Molecular ionization and dissociation also exhibit CEP dependence. It is anticipated that it may be possible to achieve specific bond breakages within a molecule by controlling the CEP of the laser pulses used to

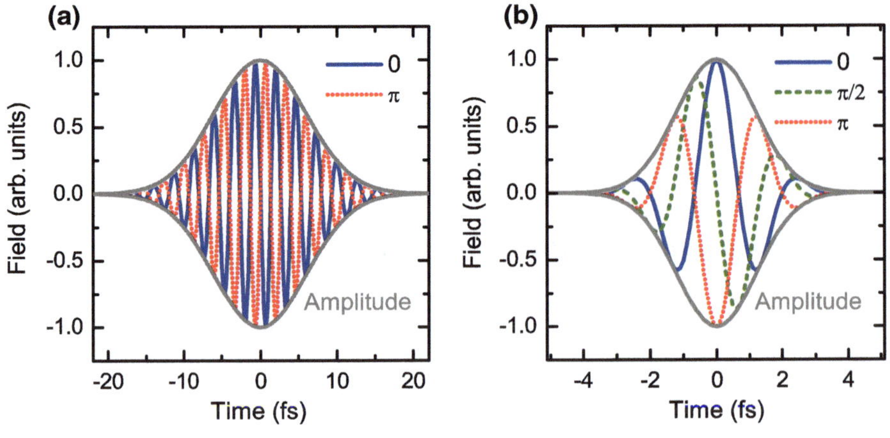

Fig. 2.6 **a** The temporal electric field (real part) of two multi-cycle pulses with CEP equal to 0 and π. **b** The temporal electric field (real part) of three single-cycle pulses with CEP equal to 0, $\frac{\pi}{2}$ and π. For the same amplitude profile, the field distribution of single-cycle pulse strongly depends on the CEP

irradiate the molecule [49]. Generally, ultrafast lasers emit pulses with random CEP values. Therefore CEP stabilization within or outside the laser is necessary for CEP sensitive experiments.

The second term in the Taylor series, φ_1 corresponds to a shift in the frequency or time (delay). Most often the physical effects and experiments are sensitive to the shape of the pulse and the magnitude of the electric field, not to the arrival time of the pulse. Occasionally, the delay is of interest in interferometric methods, in which high phase stability needs to be maintained [50, 51].

The third term (φ_2), proportional to $\frac{d^2n}{d\omega^2}$ or the group velocity dispersion (GVD), represents a linear ramp of frequency as a function of time during the pulse duration (referred to as linear chirp). Unlike the condition shown in Fig. 2.6a, the fringe spacing either increases or decreases linearly with time.

The linear optical response of the biologically relevant medium, water, is shown in Fig. 2.7a. The refractive index n is calculated using the Sellmeier equation, whereas absorbance is a measured quantity. Water is transparent up to $\sim$1000 nm. At longer wavelengths, the absorption increases rapidly, resulting in anomalous as well as higher-order dispersion [37]. Calculated second- (k'') and third- (k''') order GVD coefficients are shown in Fig. 2.7b. We see that there is a significant increase in k''' in the wavelength range above 1000 nm. Effects of k'' and k''' on the spectral phase and the intensity envelope of a Gaussian pulse centred at 1300 nm after propagating over 2 mm through water are shown in Fig. 2.7c, d, respectively. As $\frac{dn}{d\omega} > 0$ for normal dispersion, or a transparent, spectral region ($\frac{dn}{d\omega} < 0$, for anomalous dispersion), propagation through materials causes the high (low) frequency components to be delayed compared to the low (high) frequency components. This difference in speed works towards destroying the temporal overlap among the components and, hence, results in an increase in the pulse width. For few-cycle pulses, the spectral width can be substantial, almost spanning an octave. Therefore, even propagation through air can lead to significant pulse broadening. Maintaining minimum pulse duration is often mandatory as it governs the temporal resolution in many experiments, yet it is challenging, particularly for sub-50 fs pulses. Hence, proper GVD compensation and special optics (broadband and all-reflective) have to be used to minimize phase distortions.

Some materials exhibit higher-order dependence of their refractive index on frequency and, so, they also induce higher-order chirp [1]. The fourth term—proportional to $\left(\frac{d^3n}{d\omega^3}\right)$—implies a quadratic chirp wherein the central frequency arrives first while frequencies on either side arrive later. These two slightly different frequencies generate beats as a function of time, causing the pulse with cubic phase distortion to have oscillations after (or before, if the quadratic chirp is negative) the main pulse (Fig. 2.7d). Further higher-order terms yield additional distortions giving rise to extremely complex pulse profiles. For certain experiments, like SC generation, photonic crystal fibres with controlled higher-order GVD are used to optimize the spectral content.

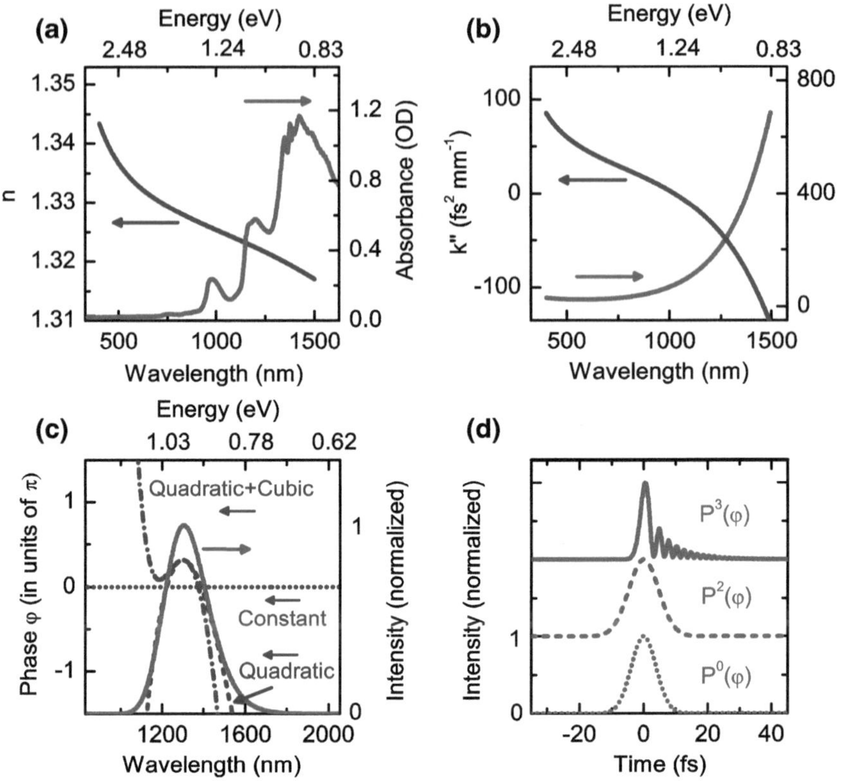

Fig. 2.7 **a** Calculated refractive index (n) and measured absorbance in water. **b** Second- (k'') and third- (k''') order GVD coefficients calculated using n shown in (**a**). The GVD in water changes sign at ~1000 nm. For wavelengths longer than 1000 nm, the anomalous chromatic regime is characterized by rapidly increasing k''' due to increasing absorbance. **c** Normalized intensity spectrum of a transform limited 10 fs Gaussian pulse centred at 1300 nm, (FWHM of ~250 nm), and its spectral phase in the presence of (i) no dispersion (Constant, dotted), (ii) only second-order GVD (Quadratic, dashed), and (iii) second- and third-order GVD (Quadratic+Cubic, solid) after traveling 2 mm in water. **d** Normalized temporal intensity profiles of the 10 fs Gaussian pulse shown in (**c**) considering GVD of different orders. $P^i(\varphi)$ represents the order of the spectral phase polynomial used to obtain the intensity profile. The temporal profiles are shifted vertically for clarity. Since Fourier transform is a linear transformation, the order of the polynomial term is maintained while transforming from the spectral to the temporal domain

2.2 Experimental Aspects

In the following sections, we discuss some methods by which ultrashort pulses are generated, characterized, and the dispersion effects are minimized. We only provide a brief and simple review of some of the topics. More details can be found in texts on ultrashort lasers and spectroscopy [1, 23, 52–56].

2.2.1 Generation of Ultrashort Pulses

Selection of the type of laser and the method of ultrashort pulse generation depends on requirements vis-à-vis the duration, energy, repetition rate and frequency [52, 53]. A short or ultrashort optical pulse can, in principle, be synthesized by starting with a continuous light source and using a fast modulator, which lets the light pass only for a short period of time. However, in such a method, the pulse duration is limited by the speed (bandwidth) of the modulator. High energy pulses with much shorter durations, in the nanosecond regime, can be generated using techniques like Q-switching, cavity dumping and gain switching.

The synthesis of a periodic pulse train can also be considered as a coherent superposition of several equally spaced frequency components (laser cavity modes). The larger the number of modes involved, the shorter the minimum pulse duration that can be achieved. A typical ultrashort laser with a cavity length of, say, 2 m, has $\sim 10^6$ modes being amplified to be able to generate a 10 fs pulse. An important aspect is that there must be a fixed phase relationship between these modes so that it is only at regular temporal positions that the electric fields of all frequency components add up (they interfere constructively) to a maximum of the total field strength. A pulse generation technique based on stabilization of the phase difference between modes is called mode locking.

In mode-locked lasers, the resonator contains an active or a passive element that facilitates the phase relationship and formation of an ultrashort pulse. The design of a mode-locked laser is generally a non-trivial task; it involves a complicated interplay of many effects, including dispersion and several nonlinear processes. Technological advancements in the last decade have led to the development of user-friendly mode-locked femtoscond oscillators, including those generating few-cycle pulses with MHz or GHz pulse repetition rates and moderate pulse energies (typically pJ to nJ). The ultrafast oscillators generate seed pulses, which can be amplified optically. Much higher pulse energies (up to several Joules) combined with lower repetition rates (up to several kHz) can be produced with one or more stages of regenerative amplifiers employing chirped-pulse amplification (CPA) that is schematically depicted in Fig. 2.8. It is also possible to reduce the repetition rate and/or to stabilize CEP either within the laser or externally by using nonlinear optical techniques. The pulse train in the MHz range is also useful for generating a stabilized frequency comb for sensitive spectral measurements discussed in Chap. 11. Several types of ultrafast lasers are commercially available: solid state (Neodymium or ytterbium-doped glasses and Titanium-sapphire lasers), fibre and semiconductor lasers. The key features for ultrafast lasers include pulse-duration, CEP stability and control, pulse-energy, repetition rate, TBP, tunability, beam profile, power and pointing stability, and ease of operation and, of course, cost.

Titanium-sapphire (Ti:sapphire, Ti:Al_2O_3) oscillators are most frequently used to generate pulses down to few femtoseconds with average powers between 100 mW to 1 W range and $\sim$80 MHz pulse repetition rate [57–59]. The gain medium comprises an Al_2O_3 matrix that is doped with Ti ions. It is optically pumped in the green region (usually with 532 nm laser light from a Nd:YAG or Nd:YVO_4 laser) and emits wavelength in the range of ~650–1100 nm. It is usually passively mode-locked, in the form of Kerr lens mode locking. A pulse duration of ≤50 fs is readily achieved. Further improvement in the laser resonator design and intra-cavity dispersion compensation has led to the development of even turn-key ultrafast lasers that generate sub-10 fs pulses.

Ti:sapphire lasers are also used in multi-pass and regenerative amplifier systems, particularly with chirped-pulse amplification [60]. They can reach enormous output peak powers of tens of TW and, in large facilities, even PW. Such high powers are essential for inducing extreme nonlinear effects like HHG and for studies of laser-plasma interactions. Nonlinear frequency conversion can also be used to further extend the emission range of a Ti:sapphire laser. Though such lasers have proved to be the workhorses of most experiments discussed in this book, recent developments of fibre and semiconductor lasers operating in near-IR spectral region have also started yielding pulses in the ~100 fs range with substantial pulse energy [53].

2.2.2 *Broadband Ultrafast Sources*

An ultrafast laser can have an octave spanning bandwidth which, coupled with SHG, can span the spectral range from the UV to the near-IR. However, the pulse energy from an oscillator is rather low (pJ) and there is no tunability. Optical parametric amplification in a nonlinear crystal like β-BaB_2O_4 (β-BBO), is particularly attractive for the generation of frequencies which are very hard to access directly. Unlike laser oscillators, optical parametric amplifiers (OPAs) provide moderate energy pulses (μJ) with tuning over a very wide spectral range [61]. They are usually pumped by the output from a regenerative amplifier (at the fundamental or the second harmonic frequency). The seed pulse needed for amplification is generally produced by SC generation in a sapphire plate. By combining multiple stages of OPAs and frequency doubling, commercial systems routinely generate sub-30 fs pulses from the UV to the mid-IR. Intense pulses of wavelength in the 1000–2500 nm range, obtained from an OPA, have found utility in probing double strand breakages in DNA, as will be discussed in Chap. 6.

Optimized SC generation is a process where the pump frequency is converted to a very broad spectral bandwidth with high spatial coherence [38, 39, 41, 42]. The pulse duration in this case is a few picoseconds and pulse energy is usually much lower than what is available from an OPA. The spectral broadening is accomplished by propagating intense ultrashort pulses through an optical fibre having high nonlinearity and waveguide structure to ensures good beam quality. Of special interest are photonic crystal fibres, mainly due to their unusual chromatic dispersion characteristics, which allow substantial broadening (spanning more than two octaves). Applications of such SC sources in ultrafast biophotonics include, but are not limited to, interferometry (Chap. 11) and spectrally- as well as temporally-resolved fluorescence microscopy (Chaps. 3, 4).

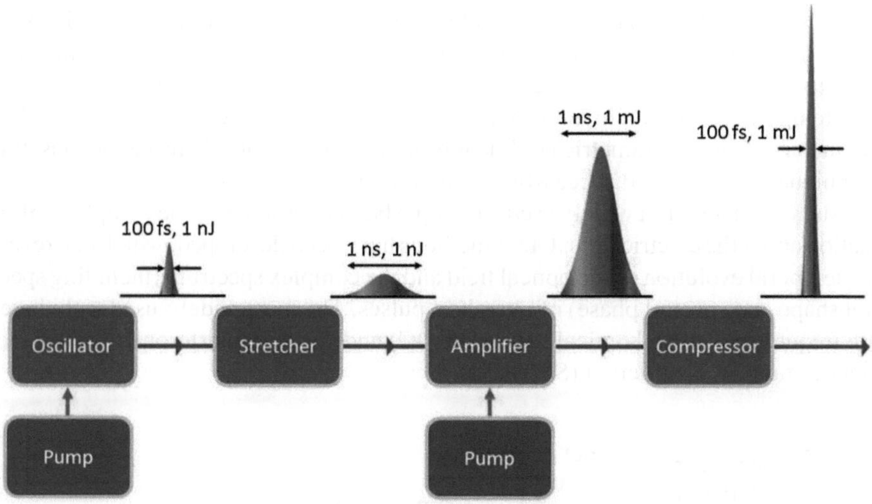

Fig. 2.8 Schematic of a typical chirped-pulse amplifier. A relatively low energy ultrashort pulse from an oscillator is stretched to ~1 ns duration by introducing chirp and then amplified to ~1 mJ. Later it is compressed back to ~100 fs. The advantage of chirp-pulse amplification is that the lower peak intensities at the amplification stage prevent breakdown of the amplifier crystal. Multiple amplification stages are employed to generate pulse energies up to several Joules

2.2.3 *Characterization of Ultrashort Pulses*

So far, we have only discussed how to generate ultrashort pulses, not how to characterize them. Measurement of pulse duration on the femtosecond timescale is challenging since the speed required is considerably faster than existing photodetectors and what can be visualized on oscilloscopes. We shall cover here a few measurement techniques capable of resolving femtosecond pulses. The techniques rely on the high speed of light and the concomitant availability of high spatial resolution, making use of the pulse itself to measure its duration [1]. We begin by discussing correlation measurements, which are easy to implement but offer only partial information. We then describe approaches that can reveal complete intensity and phase information.

Optical pulses, and pulse trains, can be characterized in terms of various parameters such as the pulse-duration, repetition rate, energy, CEP, spectrum, chirp, spatio-temporal pulse shape, and coherence. Some of these parameters—like repetition rate, pulse energy, and spectrum—are not too difficult to measure, whereas pulse duration, CEP, chirp and temporal shape are more challenging. Pulse durations down to roughly 10 ps can be measured with fast photodiodes in combination with fast sampling oscilloscopes. For the measurement of shorter pulse durations, up to fraction of a picosecond, streak cameras can be used. For even shorter pulses, techniques like optical autocorrelation become mandatory [1]. A typical set-up for autocorrelation is shown in Fig. 2.8a. It involves the overlapping of two pulse replicas in a nonlinear medium (generally a SHG crystal) and recording the generated nonlinear signal as a function of the delay between the two replicas. Typically, a collinear geometry is used to avoid loss of temporal resolution via geometric effects (finite beam angles). The temporal resolution in autocorrelation is limited by the response of the nonlinear material. Since parametric nonlinear processes are essentially instantaneous, this technique has been used successfully even for single-cycle pulses.

Autocorrelation is a widely used technique but it does not provide complete information about the electric field. Other methods have been developed, which can reveal the temporal evolution of the optical field and the complex spectrum (including spectral shape and spectral phase) of ultrashort pulses. The most widely-used techniques are frequency-resolved optical gating (FROG) and spectral interferometry for direct electric-field reconstruction (SPIDER).

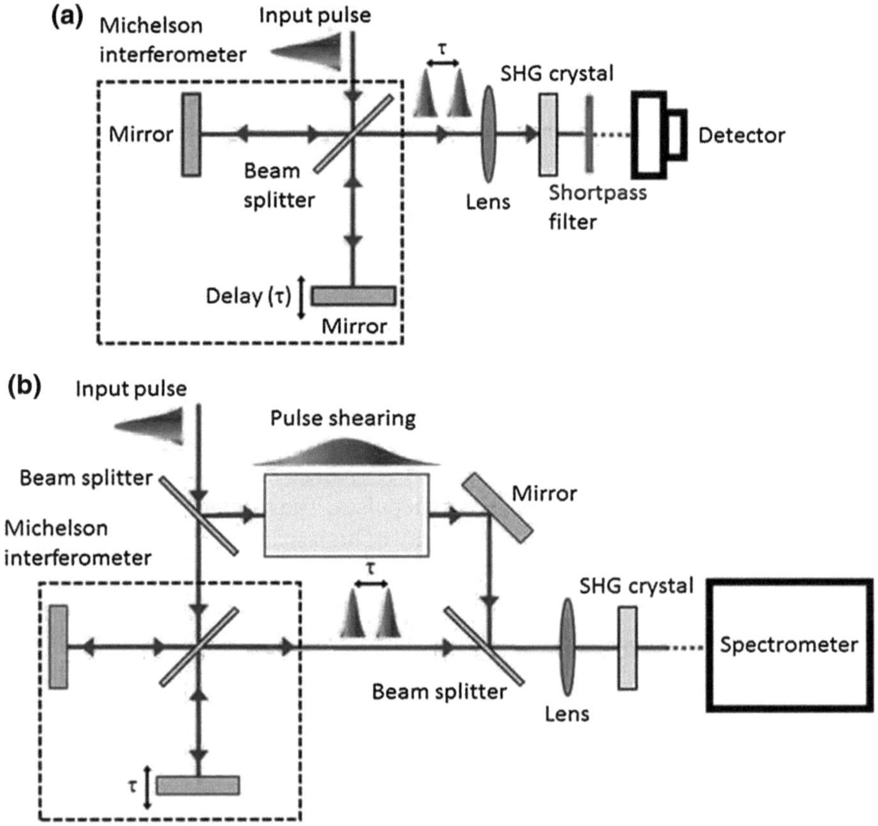

Fig. 2.9 Schematics of **a** an interferometric autocorrelator and **b** a spectral interferometry for direct electric-field reconstruction (SPIDER). The autocorrelator set-up can also be used to perform frequency resolved optical gating (FROG) by replacing the detector by a spectrometer

The FROG technique usually employs a set-up that is similar to that for an autocorrelator (Fig. 2.9a) except that the photodetector is replaced by a spectrometer [1]. The measurement involves recording a large number of spectra as a function of delay to construct a FROG trace. An iterative phase retrieval algorithm is then implemented to reconstruct the pulse shape from the FROG trace. There are different versions of FROG, each relying on different third-order nonlinear correlation processes, like nonlinear polarization, self-diffraction and transient gratings. The major disadvantage of this technique is the requirement of a phase-retrieval algorithm which may not provide a unique solution, or may not converge for a complicated pulse structure.

An alternative to FROG is spectral interferometry [62], whose major advantages are: (i) the need for a nonlinear medium is obviated and, hence, it is easy to implement, and (ii) it does not need a phase-retrieval algorithm.

The basic principle of spectral interferometry is to combine two temporally delayed pulses in a spectrometer. The optical spectrum of the superposition consists of interference fringes whose separation is governed by the relative delay and the relative chirp. Hence, the fringe spacing measured as a function of frequency yields information about the spectral phase. Along with the spectrum, the spectral phase can be used to construct the complete complex optical field in the frequency domain, which can then be Fourier transformed to obtain the electric field in the time domain. This technique works successfully provided one of the two pulses, known as the reference pulse, is already fully characterized. In case the reference pulse is not available, it has to be generated. One of the ways to generate a reference is by spectral shearing interferometry, as in SPIDER [63, 64].

A SPIDER set-up is shown in Fig. 2.9b. Here, the signal pulse is split into two replicas with a large enough temporal delay to ensure that there is no pulse overlap. A third pulse, derived from the same input pulse, is strongly temporally sheared (broadened) by sending it through a highly dispersive optical element, such as a block of glass. The long chirped pulse and the two replicas of the signal pulse are then combined in an SHG crystal. The two signal pulses overlap with different temporal portions of the chirped pulse, having different optical frequencies. Therefore the up-converted pulses also have spectral shear. The analysis of the optical spectrum of the nonlinear signal reveals the spectral phase from which the complete electric field can be constructed. SPIDER does not require a sophisticated iterative algorithm for retrieving the spectral phase. Its simple algorithm can be executed very rapidly on a laptop, allowing for fast update rates, limited only by the speed of the spectrometer. The major disadvantages of SPIDER are: (i) its complex set-up, (ii) weak up-converted signals, a consequence of the temporally sheared pulse having very low intensity, and (iii) chirp calibration.

A comprehensive comparison of different techniques is difficult because there are many variants of each technique, each developed for specific applications and each offering specific advantages under certain experimental circumstances. Several commercial pulse characterization systems are available which can be used for a wide range of ultrashort pulses.

2.2.4 Dispersion Compensation

Control of dispersion is the key to obtaining higher temporal resolution when using ultrashort pulses. Dispersion compensation essentially means cancelling the chromatic dispersion of some optical element(s). However, the term is often used in a more general sense, meaning the control of the overall phase distortions within the laser and associated optical system. The main goal is to avoid pulse distortion and broadening. Several techniques have been developed for dispersion control. These can be grouped into two categories:

(i) *Linear pulse compression*: For chirped pulses, the duration can be reduced by compensating the chirp, in other words, by "flattening" the spectral phase. Since the material dispersion introduces positive chirp, the compensation can be accomplished by sending the pulses through a set of optical elements which introduce negative chirp. The aim is to delay the lower frequency components with respect to the higher frequency components so as to compensate the material dispersion. Optical configurations like a prism pair (a prism compressor) [65, 66], grating pair (a grating compressor) [67], dispersion compensating optical fibre [68], chirped mirrors [69], chirped Bragg gratings or a liquid crystal array have all been used. Typical prism and grating pair compressor designs shown in Fig. 2.10 are often used due to their simplicity and tunability. Chirped mirrors capable of compensating higher order chirp, for even sub-10 fs pulses, have also been developed [58], which are much easier to integrate within a set-up, but there is no control over the amount of

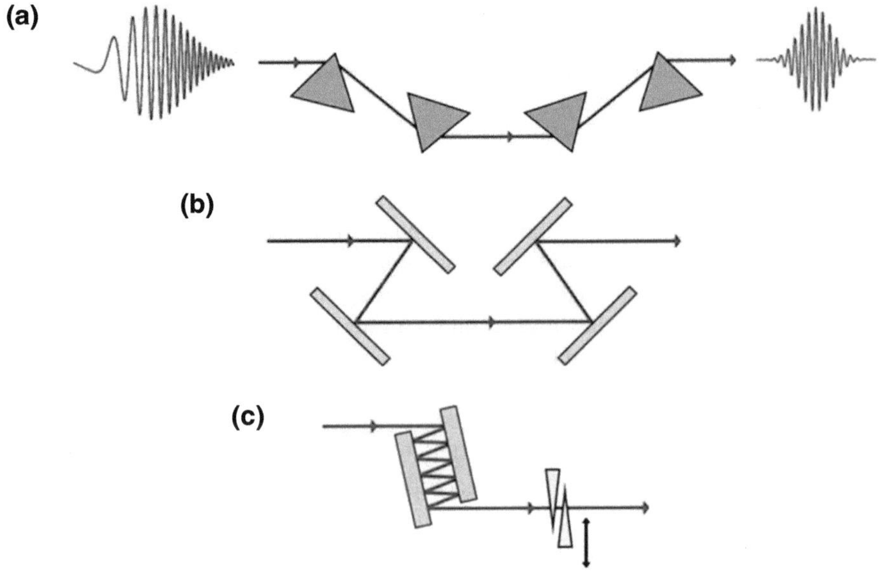

Fig. 2.10 Dispersion control of ultrashort pulses by **a** a prism compressor, **b** a grating compressor, and **c** a set of chirped mirrors and wedges

chirp introduced unless a pair of fused silica or CaF_2 wedges are used for fine adjustment (Fig. 2.10). In linear compressor configurations, only the phase of individual frequency component is altered. There is no change in the spectral bandwidth of the pulse assuming there are no absorption losses. Therefore the minimum attainable pulse duration is set by the optical bandwidth of the pulses. Under ideal conditions, a transform limited pulse is obtained.

(ii) *Nonlinear pulse compression*: In a first step, the optical bandwidth is increased, typically with a nonlinear interaction such as self-phase modulation in an optical fibre. In most cases, this leads to chirped pulses, with a duration longer than the original pulse. Thereafter, the pulse duration can be made considerably shorter by linear (dispersive) compression. Nonlinearity can also be used to compensate the material dispersion, for example, by propagation through a hollow-core fibre filled with a rare gas (typically Ne), a technique that is now routinely used for generating single- and sub-cycle pulses [70, 71] or via a filamentation based technique [72].

Dispersion control certainly helps in reducing the chirp but it is difficult to totally eliminate it; higher-order phase distortions are, almost always, found to persist. Therefore, as far as possible, experiments using ultrashort pulses necessitate the use of customised broadband and all-reflective optics to ensure minimum pulse chirping and broadening [73].

References

1. R. Trebino, *Frequency Resolved Optical Gating: The Measurement of Ultrashort Pulses* (Kluwer Academic Publishers, USA, 2000)
2. M. Wegener, *Extreme Nonlinear Optics* (Springer, Berlin Heidelberg, 2005)
3. J.D. Jackson, *Classical Electrodynamics*, 3rd edn. (Wiley, Singapore, 2005)
4. L.D. Landau, L.P. Pitaevskii, E.M. Lifshitz, *Electrodynamics of Continuous Media*, 3rd edn. (Butterworth-Heineman, Oxford, 2004)
5. D.M. Volkov, Z. Phys. **94**, 250 (1935)
6. N.D. Sengupta, Bull. Math. Soc. (Calcutta) **41**, 187 (1941)
7. N.D. Sengupta, Bull. Math. Soc. (Calcutta) **44**, 175 (1952)
8. Vachaspati, Phys. Rev. **128**, 664 (1962)
9. Vachaspati, Phys. Rev. **130**, 2598(E) (1963)
10. H.R. Reiss, J. Math. Phys. **3**, 59 (1962)
11. A.I. Nikishov, V.I. Ritus, Sov. Phys. JETP **19**, 529 (1963)
12. R.W. Boyd, *Nonlinear Optics*, 3rd edn. (Academic Press, London, 2008)
13. L.V. Keldysh, Sov. Phys. JETP **20**, 1307 (1965)
14. F.H.M. Faisal, J. Phys. B **6**, L89 (1973)
15. H.R. Reiss, Phys. Rev. A **22**, 1786 (1980)
16. H. Haug, S.W. Koch, *Quantum Theory of the Optical and Electronic Properties of Semiconductors*, 4th edn. (World Scientific, Singapore, 2004)
17. Y.R. Shen, *The Principles of Nonlinear Optics* (Wiley, New York, 1984)
18. M. Bass, P.A. Franken, A.E. Hill, C.W. Peters, G. Weinreich, Phys. Rev. Lett. **8**, 18 (1962)
19. J.A. Armstrong, N. Bloembergen, J. Ducuing, P.S. Pershan, Phys. Rev. **127**, 1918 (1962)
20. P.A. Franken, A.E. Hill, C.W. Peters, G. Weinreich, Phys. Rev. Lett. **7**, 118 (1961)
21. M. Goeppert-Mayer, Ann. Phys. **9**, 273 (1931)
22. W. Kaiser, C. Garrett, Phys. Rev. Lett. **7**, 229 (1961)

23. R.R. Alfano, *The Supercontinuum Laser Source*, 2nd edn. (Springer, Heidelberg, 2006)
24. G.H.C. New, J.F. Ward, Phys. Rev. Lett. **19**, 556 (1967)
25. J.A. Squier, M. Müller, G.J. Brakenhoff, K.R. Wilson, Opt. Express **3**, 315 (1998)
26. R.Y. Chiao, E. Garmire, C.H. Towenes, Phys. Rev. Lett. **13**, 479 (1964)
27. Y.R. Shen, Prog. Quantum Electron. **4**, 1 (1975)
28. A.K. Dharmadhikari, F.A. Rajgara, D. Mathur, Appl. Phys. B **80**, 61 (2005)
29. J.S. D'Souza, J.A. Dharmadhikari, A.K. Dharmadhikari, B.J. Rao, D. Mathur, Phys. Rev. Lett. **106**, 118101 (2011)
30. A.K. Dharmadhikari, H. Bharambe, J.A. Dharmadhikari, J.S. D'Souza, D. Mathur, Phys. Rev. Lett. **112**, 138105 (2014)
31. J. Thomas, R. Bernard, K. Alti, A.K. Dharmadhikari, J.A. Dharmadhikari et al., Opt. Commun. **304**, 29 (2013)
32. S.D. George, U. Ladiwala, J. Thomas, A. Bankapur, S. Chidangil, D. Mathur, App. Surf. Sci. **305**, 375 (2014)
33. A. Couairon, A. Mysyrowicz, Phys. Rep. **441**, 47 (2007)
34. S.L. Chin, S.A. Hosseini, W. Liu, Q. Luo, F. Théberge et al., Can. J. Phys. **83**, 863 (2005)
35. R.L. Fork, C.V. Shank, C. Hirlimann, R. Yen, W.J. Tomlinson, Opt. Lett. **8**, 1 (1983)
36. P. Vasa, M. Singh, R. Bernard, A.K. Dharmadhikari, J.A. Dharmadhikari, D. Mathur, Appl. Phys. Lett. **103**, 111109 (2013)
37. P. Vasa, J.A. Dharmadhikari, A.K. Dharmadhikari, R. Sharma, M. Singh, D. Mathur, Phys. Rev. A **89**, 043834 (2014)
38. P. Panagiotopoulos, D.G. Papazoglou, A. Couairon, S. Tzortzakis, Nature Commun. **4**, 2622 (2013)
39. F. Silva, D.R. Austin, A. Thai, M. Baudisch, M. Hemmer et al., Nature Commun. **3**, 807 (2012)
40. G. Stibenz, N. Zhavoronkov, G. Steinmayer, Opt. Lett. **31**, 274 (2006)
41. S.E. Harris, A.V. Sokolov, Phys. Rev. A **55**, R4019(R) (1997)
42. H.S. Chan, Z.M. Hsieh, W.H. Liang, A.H. Kung, C.K. Lee et al., Science **331**, 1165 (2011)
43. C. Rewitz, T. Keitzl, P. Tuchscherer, J.S. Huang, P. Geisler et al., Nano Lett. **12**, 45 (2012)
44. K. Lindfors, T. Kalkbrenner, P. Stoller, V. Sandoghdar, Phys. Rev. Lett. **93**, 037401 (2004)
45. A. McPherson, G. Gibson, H. Jara, U. Johann, T.S. Luk et al., J. Opt. Soc. Am. B **4**, 595 (1987)
46. J. Seres, E. Seres, A.J. Verhoef, G. Tempea, C. Streli et al., Nature **433**, 596 (2005)
47. P.M. Paul, E.S. Toma, P. Breger, G. Mullot, F. Augé et al., Science **292**, 1689 (2001)
48. D. Mathur, K. Dota, A.K. Dharmadhikari, J.A. Dharmadhikari, Phys. Rev. Lett. **110**, 083602 (2013)
49. D. Mathur, K. Dota, D. Dey, A.K. Tiwari, J.A. Dharmadhikari et al., J. Chem. Phys. **143**, 244310 (2015)
50. S. Mukamel, Annu. Rev. Phys. Chem. **51**, 691 (2000)
51. M.H. Cho, Chem. Rev. **108**, 1331 (2008)
52. K. Wolfgang, *Ultrashort Laser Pulses: Generation and Applications* (Springer, Germany, 1993)
53. U. Keller, Nature **424**, 831 (2003)
54. G.R. Fleming, *Chemical Applications of Ultrafast Spectroscopy* (Oxford University Press, New York, 1986)
55. O. Svelto, S. De Silvestri, G. Denardo, *Ultrafast Processes in Spectroscopy* (Plenum Press, New York, 1996)
56. M. Braun, P. Gilch, W. Zinth, *Ultrashort Laser Pulses in Biology and Medicine* (Springer, Berlin, 2008)
57. P.F. Moulton, J. Opt. Soc. Am. B **1**, 125 (1986)
58. A. Stingl, M. Lenzner, Ch. Spielmann, F. Krausz, R. Szipöcs, Opt. Lett. **20**, 602 (1995)
59. U. Morgner, F.X. Kärtner, S.H. Cho, Y. Chen, H.A. Haus et al., Opt. Lett. **24**, 411 (1999)
60. D. Strickland, G. Mourou, Opt. Commun. **56**, 219 (1985)
61. G. Cerullo, S. De Silvestri, Rev. Sci. Instrum. **74**, 1 (2003)
62. V. Wong, I.A. Walmsley, Opt. Lett. **19**, 287 (1994)
63. C. Iaconis, I.A. Walmsley, Opt. Lett. **23**, 792 (1998)

64. L. Gallmann, D.H. Sutter, N. Matuschek, G. Steinmeyer, U. Keller, Appl. Phys. B **70**, S67 (2000)
65. O.E. Martinez, J.P. Gordon, R.L. Fork, J. Opt. Soc. Am. A **1**, 1003 (1984)
66. R.L. Fork, O.E. Martinez, J.P. Gordon, Opt. Lett. **9**, 150 (1984)
67. C.V. Shank, R.L. Fork, R. Yen, R.H. Stolen, W.J. Tomlinson, Appl. Phys. Lett. **40**, 761 (1982)
68. H.E. Bates, J. Opt. Soc. Am. **70**, 1017 (1980)
69. E.J. Mayer, J. Möbius, A. Euteneuer, W.W. Rühle, R. Szipőcs, Opt. Lett. **22**, 528 (1997)
70. M. Nisoli, S. De Silvestri, O. Svelto, R. Szipőcs, K. Ferencz et al., Opt. Lett. **22**, 522 (1997)
71. J. Biegert, J.-C. Diels, J. Opt. Soc. Am. B **18**, 1218 (2001)
72. A.K. Dharmadhikari, J.A. Dharmadhikari, F.A. Rajgara, D. Mathur, Opt. Express **16**, 7083 (2008)
73. B. Piglosiewicz, D. Sadiq, M. Mascheck, S. Schmidt, M. Silies et al., Opt. Express **19**, 14451 (2011)

Chapter 3
Nonlinear Microscopy

Abstract An overview is presented of the application of ultrafast nonlinear optics to biophotonic microscopy, with the essential physics being based on multiphoton excitation. The use of a femtosecond laser permits high enough incident power levels for nonlinear effects to occur efficiently while, at the same time, being low enough to ensure sample integrity. Furthermore, the use of infrared light enables greater depth penetration, an important consideration in biophotonic applications. We also discuss successes of super-resolution microscopy, which is founded essentially on the "manipulation" of molecular spectra, and the anticipated successes of 4D electron microscopy in unravelling ultrafast biological phenomena.

3.1 Visualization of Biological Processes and Entities

Physical scientists are most familiar with experimental inquiry into "structure" being the primary driver that leads to the formulation and development of deeper insights into "processes" and "phenomena". In the life sciences also, structural studies have acted as significant precursors of many contemporary developments. Here, Photonics continues to play an exemplary role by providing tools like X-ray spectroscopy, synchrotron radiation sources and, now, free-electron-laser based sources of ultrafast X-ray pulses. Structure determination again remains the forerunner in the quest for tangible links between the structure of biological entities and their biological function. The domain of the life sciences now seems to be organized according to the following hierarchy:

$$Gene/DNA \leftrightarrow proteins \leftrightarrow organelles \leftrightarrow cells \leftrightarrow tissues \leftrightarrow organs \leftrightarrow organisms.$$

From a systems viewpoint, it might be expected that structure-function relationships would ultimately address the highest—most difficult—stage of such a hierarchy. However, it seems that progress in even the first few steps in this hierarchy continue to be both challenging and a source of inspiration for those engaged in biophotonics.

A single cell maintained under physiological conditions serves to provide the biological link between molecular level details that are becoming experimentally accessible and the systems level details that life scientists strive for. Theoretical

P. Vasa and D. Mathur, *Ultrafast Biophotonics*, Biological and Medical Physics, Biomedical Engineering, DOI 10.1007/978-3-319-39614-9_3

work based on physics methodologies, especially those that relate to whole cell simulations, are beginning to find utility in providing linkages to experiments of the type we shall discuss in this chapter.

On the experimental front, super-resolution imaging of cell structures has started offering exciting prospects of new biological insights being developed; indeed, a so-called skeleton of neurons has already been discovered [1] and time-resolved dynamics in cell membranes have been carried out [2]. Such experiments, accompanied by appropriate computer simulations, may well pave the way to the development of methods that will enable in vivo diagnostics, as well as real-time probing and characterization of cellular machinery and motors. The embracing of photonics techniques by the neurobiology and cell development communities is likely to benefit both communities. There is now considerable interest in probing molecular machinery using fluorescence labelling, including application of in vivo single-molecule experiments based on femtosecond lasers. Single-molecule sensitivity is now beginning to be a reality by employing photonics tools like high-resolution laser spectroscopy, total internal reflection, and photo-localization [3]. This is remarkable considering the intrinsically "noisy" background that pervades a living cell's environment. There are sound grounds for optimism that results forthcoming from single molecule spectroscopy and dynamics experiments (Chap. 4) will help develop hitherto-unavailable insights into complex "real" experiments of interest and relevance to the life sciences community. It is, therefore, of interest to provide an overview of some pertinent recent developments.

3.2 Multiphoton Microscopy

Conventional microscopy has long been the workhorse of experimental biology—in both its variants: brightfield microscopy and darkfield microscopy. In brightfield microscopy, the sample to be viewed is illuminated by white light, and contrast is obtained because of differential absorbance of the light transmitted through the sample. In contrast, darkfield microscopy relies on collection of only the light that is scattered off the sample: it rejects the transmitted light. As a consequence, enhanced contrast is obtained of unstained and live biological entities. Both variants suffer the constraint that is imposed by the expectedly shallow penetration of the incident light into the sample. Hence, it is mostly thin samples that can be imaged.

Confocal microscopy seeks a way around this constraint by rejecting light that emanates from planes above and below the focal plane (where the sample is located). It does so by using a focused laser beam to scan the sample and by placing a pinhole in front of the photon detector. Of course, the pinhole physically restricts the flux of photons reaching the detector and, consequently, limits the ultimate sensitivity that can realistically be expected.

A totally alternative approach emerged in 1990 when it was shown that multiphoton techniques could profitably be incorporated into microscopy [4]. It was two-photon excitation that was used to demonstrate 3D imaging. Since then, multiphoton excitation has become the norm; such microscopy largely relies on high repetition rate femtosecond lasers. Although the peak powers that are delivered by femtosecond lasers can be very high, the low duty cycle helps minimize the average power that is deposited into an irradiated bio-sample while, maximizing the efficiency for multiphoton excitation. At the same time, a built-in ("free") advantage is obtained: confocality. In order to appreciate this "free" advantage consider that, for a spatially uniform sample, single-photon excitation would yield fluorescence signals generated across the entire Rayleigh range. This would be the case equally from each z-plane below and above the focal plane where the sample is located. In contrast, multiphoton excitation automatically provides sharper depth discrimination: this is because now the major portion of the fluorescence signal originates from a much smaller focal volume due to the z^{-2} dependence of excitation probability. Significant fluorescence signals are generated only from around the focal volume where the laser intensity (the photon density) is high enough for multiphoton events to occur.

By way of illustration, consider excitation by 960 nm laser light that is focused by an objective lens with numerical aperture (NA) of 1.25. The spread function for two-photon excited emission would, typically, have a full width at half maximum (FWHM) of only ~150 nm (compared to ~300 nm at 960 nm) in the radial direction; the corresponding FWHM in the axial direction would be ~400 nm (~900 nm at 960 nm). A self-explanatory schematic illustration is shown in Fig. 3.1 which confirms that a much smaller focal volume participates in multiphoton events compared to single photon events.

Long wavelength light is usually preferred for two-photon microscopy as it provides the dual advantage of (i) relatively low linear absorption and (ii) less scattering. Both of these are particularly advantageous in biophotonics because they facilitate deeper penetration into a bio-sample: relatively thick bio-samples can be readily imaged. Longer wavelength light also has the additional advantage of reduced propensity for inducing photodamage. This is important as longer imaging times thus

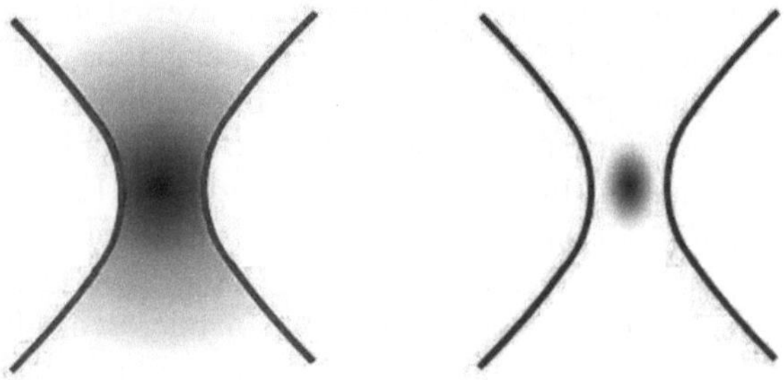

Fig. 3.1 The effective interaction volume for (*left image*) single photon and (*right image*) multiphoton excitation. Significant fluorescence signals are generated only from within the interaction volume. Note the very significantly truncated interaction volume that arises in the case of multiphoton excitation

become feasible, a specially important consideration when many scans need to be made for construction of 3D images. Are there potential disadvantages to using long wavelengths? There is possibly a larger propensity of photobleaching which, for two-photon microscopy, depends on the intensity that is incident at the focal plane [5]. In contrast, photobleaching in the case of single photon excitation is predominantly affected by the wavelength, not so much by the intensity, of the illuminating light. Other potential disadvantages involve infrared activation of certain fluorophores. Such activation gives rise to O-atoms being produced which have been shown to trigger successive cascades of biochemical damage in cells [6]. It is important to note in this context that in plant tissues, the photosynthesis machinery actually absorbs near infrared light [7].

It will be readily appreciated that two-photon excitation processes differ from one-photon ones from a quantum mechanical viewpoint. One obvious difference is that a different set of selection rules come into play in the two cases. Consequently, it is usually not practicable to make use of a single-photon excitation spectrum to derive the two-photon analogue. This has implications in the practical implementation of multiphoton microscopy: wavelength scans become necessary if a 2-photon absorption spectrum of each protein or fluorescent dye is to be derived [8]. At the same time, there is an on-going need to discover, synthesize, and characterize new fluorescent proteins that possess the requisite two-photon properties to make them suitable for imaging [9–12]. There has been considerable contemporary interest in developing infra-red fluorescent proteins [13] that make possible imaging from deeper within tissues. Apart from proteins, other (non-biological) fluorescent molecules (usually organic molecules), atoms, and atomic ions (both positively and negatively charged ones) continue to be explored for possible utility in multiphoton imaging. The main driving force behind these efforts is the fact that such probes can be directly used on bio-samples, without recourse to prior genetic engineering. Recent examples of such developments have involved metal ions like Ca^{2+}, Mg^{2+}, Zn^{2+} [14, 15], O- and N-atoms [16, 17], organic molecules such as saccharides [18], and negatively charged ions [19].

So, how is the influence of multiphoton microscopy being felt in biophotonics? It appears that influences are beginning to be exerted in a number of areas, ranging from physiology and embryology to tissue engineering and neurobiology. There are already some specific areas where it has been shown that two-photon microscopy scores over conventional confocal microscopy, for instance in the development of model organisms like *Drosophila melanogaster* [20], *Danio renio* [21–23], and *Caenorhabditis elegant* [24, 25]. All three species are known to be readily susceptible to photodamage, thereby precluding employment of conventional confocal microscopy techniques. In contrast, two-photon microscopy has made all three species amenable to imaging, with adequate resolution and sensitivity at various depths within the tissue. Thus, two-photon microscopy is beginning to facilitate direct visualization of cells and cellular dynamics in physiologically-relevant environments.

Time-dependent studies have also begun to be performed, making feasible in vivo probing of real-time changes of structure and function at microscopic levels and, indeed, in live animals [26, 27]. Extension to clinical applications bring to the fore the problem of attaining high-speed imaging deep into tissues. Some of the associated technical challenges are being addressed in studies of human skin [28].

A cogent review has recently appeared that details the techniques of two-photon and multiphoton microscopy, with discussion on how these are beginning to be applied by neurobiologists to gain insights into learning and memory [29]. In the following we discuss just one of several applications of multiphoton microscopy in contemporary, real-life biophotonics.

The process of amyloid fibrillation involves the misfolding of native soluble proteins into fibrils that are insoluble (see Fig. 3.2). Such fibrils are usually composed of cross-β-sheets although, in some systems (like α-synuclein), the β-sheets can co-exist with α-helices in equilibrium. In a general sense, fibrils are of considerable interest in themselves as they can be thought of as one-dimensional entities: they are typically several micrometres in length while their diameter is only of the order of 10 nm. The fibrillization process is initiated when an initially unfolded protein undergoes misfolding. The importance of protein fibrillization from the biophotonics perspective lies in the potential involvement of amyloid fibrils in a large number of amyloidogenic diseases like Parkinson's disease, Alzheimer's disease (AD), and prion-associated encephalopathies. It has been found that all these share fibril formation as the common symptom [30]. It has recently been discovered that amyloid fibrils possess the somewhat unexpected but most useful property of having very large cross sections for multiphoton absorption [31]; such enhanced nonlinear absorption is not observed in non-fibrillized proteins and provides a ready diagnostic for detection of fibrils. Nonlinear optical experiments, including Z-scan and pump-probe measurements, have indicated that α-synuclein fibres, lysozome and insulin β-amyloids all exhibit high propensities for 2-photon, 3-photon, and multiphoton absorption [31], with the origin of such enhanced nonlinear absorption being ascribed to efficient through-space dipolar coupling between excited electronic states of aromatic amino acids that are densely packed within the fibrillized proteins [32, 33].

While the presence of amyloid plaques is known to be strongly correlated with disease, it has, for many years, remained unclear whether fibrils themselves might be the causal agents of diseases [34–36]. Neuropsychological analysis of patient's clinical symptoms include assessment of memory impairment and deficiencies in cognition as well as post-mortem examination of brain tissue. Unfortunately, it remained the case for many years that AD could only be diagnosed with certainty after death. Even now the most definitive diagnostics continue to depend on analysis of postmortem tissue from AD-afflicted patients; such analysis invariably reveals the existence of extensive neurofibrillary tangles and plaques of amyloid-beta (Aβ). Hence, plaque pathology is of prime importance from the viewpoint of diagnostics and has, for many years now, relied on histological methods. The advent of multiphoton microscopy has opened new prospects of in vivo imaging of Aβ plaques in intact brains, offering the ability to achieve minimally invasive early diagnosis of AD. Competing contemporary methods of such imaging include positron emission tomography (PET), magnetic resonance imaging (MRI), and Near-infrared fluorescence imaging (NIR). It is, therefore, of interest to take cognizance of possible advantages that multiphoton

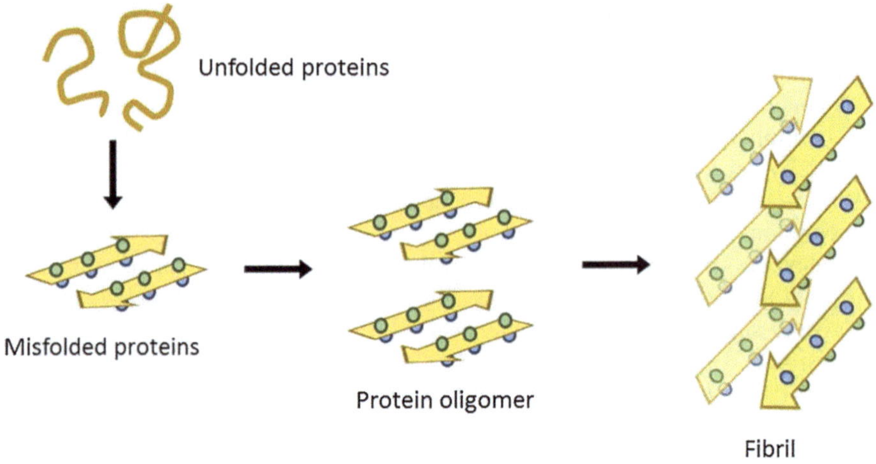

Fig. 3.2 Fibrillization of an initially misfolded protein. The initial steps wherein an unfolded protein either folds or misfolds on a potential energy landscape. Typically, fibrils comprising anti-parallel β-sheets, as shown, would have a length of several micrometres while their diameter would be only $\sim$10 nm. Some natural biopolymer aggregates comprise both the β-sheets and α-helices, in co-existence with each other. They can exhibit remarkable material properties, like pull strengths similar to those associated with steel, and mechanical shear stiffness akin to what silk possesses [33]

methods might have over these (well-established) techniques. Foremost is the important advantage of superior spatial resolution that multiphoton microscopy offers (see Table 3.1).

Several consequences follow from the superior spatial resolution that multiphoton microscopy offers [37], such as its intrinsic ability to undertake imaging of (i) a variety of objects (like dendritic spines and the post-synaptic apparatus of excitatory synapses) and (ii) dynamic processes like the onset and subsequent growth of amyloid plaques, and Ca^{2+} ion transients.

Diagnosis of AD in living patients depends on the detection of neuropathological lesions. Such detection can either be direct, via brain imaging or, as was the case in early studies, by an indirect route that involved the monitoring of a suitable biomarker that could follow the development of lesions [38, 39]. Use of an appropriate

Table 3.1 Comparison of techniques used for in vivo imaging of the amyloid protein

Technique	Spatial resolution (mm)	Sensitivity to Aβ
Multiphoton microscopy	0.001	High
Positron emission tomography	2–10	Micromolar
Magnetic resonance imaging	0.05–1.0	Millimolar
Near-infrared spectroscopy	1–4	High

biomarker was also, potentially, useful from a therapeutic viewpoint as it could be used as a monitor for the efficacy with which lesions were being removed as part of a treatment protocol. As already noted, considerable effort was expended in developing appropriate imaging agents that would enter the brain and specifically target plaques and tangles so as to facilitate direct imaging in humans. Such agents would cross the blood-brain barrier and permit administration of some amyloid-targeting reagent which could be labeled with radioactive ^{13}C for PET [40]. The half-life of ^{13}C is 20 min and, consequently, these early diagnostics relied on fast brain entry, equally fast clearance from the brain of non-specifically bound and free tracers, and long-lasting binding of the agent to Aβ plaques. Even so, as is evident from Table 3.1, the ultimate spatial resolution offered by this technique was very limited.

The advent of multiphoton microscopy altered the scenario in dramatic fashion. Bacskai and coworkers were amongst the earliest to demonstrate the utility of multiphoton microscopy by experimentally determining the real-time biodistribution kinetics of an imaging agent in transgenic mouse models of AD [39]. They used a chemical derivative of thioflavin T, known as Pittsburg compound B (PIB), that labels plaques and cerebral amyloid angiopathy in tissue sections taken from AD patients, to successfully image—with sub-μm resolution—the brains of living transgenic mice as PIB was being administered. It was discovered that PIB impregnated the mouse brain very rapidly and labeled amyloid deposits within a period of a few minutes [39]. Moreover, nonspecific binding was rapidly cleared while, at the same time, specific labeling was prolonged. Wild-type mice showed rapid brain entry and clearance of PIB without any binding, offering convincing indications that PIB might also be a good amyloid-imaging agent in humans diagnosed with AD, or those with a propensity for developing AD.

Several other imaging agents have now been developed with some, like methoxy-X04, permitting plaques in transgenic mouse models of AD to become amenable to visualization in four dimensions: in the three spatial dimensions as well as over time [37]. Such studies have brought to the fore a major strength of multiphoton microscopy: the in-built ability to repeatedly scan the same region of the brain in the same animal. These type of 4D imaging methods frequently require recourse to a second label that helps to identify the vasculature, thus providing a reference landmark that helps identify features in the three-dimensional volumetric pixel (sometime referred to in the literature as voxel). The second label is usually a fluorescent dye conjugated to dextran that is injected into the tail vein of the mouse [37].

Multiphoton microscopy has certainly enabled qualitative advances to be made in the carrying out of in vivo imaging; it has rapidly become one of the tools of choice not only in the investigation of amyloid dynamics in AD but also in experimental studies of diverse cellular interactions with amyloid aggregation. A cogent review is to hand that describes recent progress [37].

3.3 Second Harmonic Microscopy

There have been other developments in microscopy that have also relied on, or have been driven by, ultrafast techniques. These include microscopy based on photon scattering and nonlinear optical phenomena, such as second harmonic generation (SHG) microscopy [41] and coherent anti-Stokes Raman scattering microscopy [42]. Both have found considerable utility in imaging thick bio-samples using relatively low values of incident laser power.

In the case of SHG microscopy, nature has been particularly helpful in endowing several biological tissues and biological assemblies—collagen, microtubules, myosin—with very high hyperpolarizabilies or non-parabolic potential energy surfaces (Chap. 2). This makes such entities particularly useful as they can be considered to have in-built light sources. For instance, fibrillar collagen—possibly the most abundant protein to be found in mammals—possesses an extraordinarily high hyperpolarizability; its value is only about tenfold less than that of crystalline quartz [43]. Consequently, SHG signals from collagen readily yield high-contrast images of not only tissues but also of the boundaries that exist between different tissue types. This capability has enabled microscopic visualization of the pathological effects that diseases have on tissues; an example is the effect on tissues of the growth of cancerous tumours [44].

The physics that underpins SHG microscopy is somewhat more complex than what was encountered in fluorescence-based two-photon or multiphoton microscopy. In SHG microscopy, account has to be taken of a coherent optical process that demands phase matching from all parts of the nonlinear focal volume [41, 45]. The retention of phase information implies that useful information is obtained not only by measuring signal intensity but also by monitoring signal directionality, as discussed in the following. This facet of SHG microscopy is of particular importance in the case of thick biological samples, like type I collagen fibrils, whose diameters can be more or less the same as the wavelength of visible light [46]. Elegant uses of SHG microscopy, including detection of both forward-scattered and back-scattered SHG signals, have been reported in in vivo studies of tissues under dynamical conditions like the healing of wounds, development, and malignancy [44, 46, 47].

Analysis of the experimental data that emerge out of SHG microscopy appears to be a reasonably straightforward task. The output polarization of the SHG signal is measured and can usually be fitted using equations that are familiar to the photonics community. One begins by expressing the electric dipole moment per unit volume—the polarization—that is induced in an electric field **E**:

$$\mathbf{P} = \mathbf{P_0} + \alpha : \mathbf{E} + \beta : \mathbf{EE} + \gamma : \mathbf{EEE} + \cdots, \tag{3.1}$$

where P_0 denotes the static dipole density; α, β and γ are, respectively, the matrices for the polarizability, and first- and second-order hyperpolarizabilities. It may be noted that α, β, and γ are the molecular equivalents of corresponding macroscopic susceptibilities, $\chi^{(i)}$ (Chap. 2).

An incident laser field, of frequency ω at Cartesian coordinates (x, y, z), will induce a second-order polarization at frequency 2ω in the ith direction that can be represented by

$$\mathbf{P}_{2\omega,i}(x, y, z) = \sum_{j,k} \langle \beta_{ijk}(x, y, z) \rangle \mathbf{E}_{\omega,j}(x, y, z) \mathbf{E}_{\omega,k}(x, y, z). \tag{3.2}$$

This equation implies the existence of 27 possible combinations. So, if it is collagen that is being subjected to investigation, and the collagen fibre is aligned along one axis, say in the y-direction, and the incident laser polarization is also along the y-direction, then we succeed in substantially reducing the number of possible combinations to only one:

$$\mathbf{P}_{2\omega,x} = \beta_{xyy} \mathbf{E}_{\omega,y} \mathbf{E}_{\omega,y}, \tag{3.3}$$

where $\mathbf{E}_{\omega,y}$ denotes the strength of the optical field that is polarized along the y-direction. Hence, the scattered light is essentially only in the forward direction in this case.

As already noted, the SHG signal in this type of microscopy arises only from within the focal volume (Fig. 3.1). If a bio-sample of finite dimensions in the x-, y-, and z-directions acts as the scatterer, a measurement of the ratio of the forward to backward SHG signal should make it possible to assess the axial size of the location from which the scattering occurs. Using this approach, Williams and coworkers [46] have shown that SHG signals emanate not from the bulk but, essentially, from the shell of the collagen fibrils. This contradicts the long-held notion developed on the basis of years of conventional electron microscopy imaging, and leads to the suggestion that the SHG-generating shell is a physical characteristic of wet fibrils. Most interestingly, this points to the possibility of developing collagen endoscopy that will surely be of considerable use in probing physiologically-relevant structural alterations on size scales that may be smaller than the resolution limit of conventional optical microscopy.

3.4 Combination Nonlinear Microscopy

It has also become possible to combine the individual advantages of multiphoton fluorescence microscopy with second- and third-harmonic generation on the same microscope platform in order to achieve 3D image reconstructions. Imaging that relies on third-harmonic generation has been shown to be of particular utility as a non-destructive, non-invasive tool that enables in vivo cellular processes to be monitored, and this has been demonstrated in studies of neuronal degeneration [24]. It is of interest to illustrate the combined nonlinear microscopy technique by considering the relative simplicity of the experimental approach.

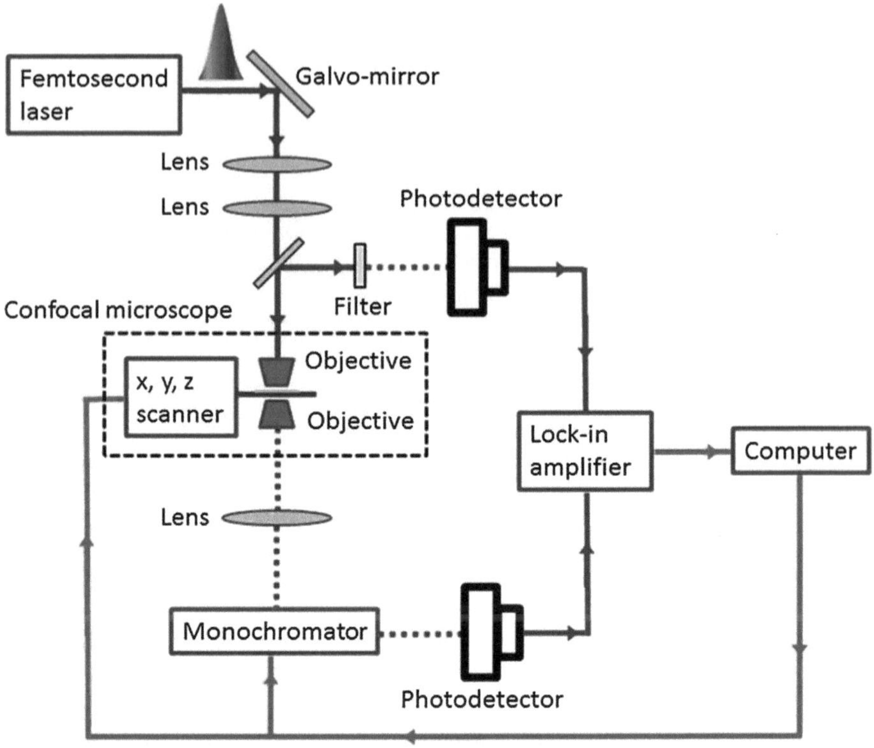

Fig. 3.3 Schematic depiction of an arrangement developed for carrying out nonlinear microscopy [24]: a combination of two-photon induced fluorescence, second-harmonic generation, and third-harmonic generation

Figure 3.3 shows a schematic outlay of one of several possible configurations that can be used for combined nonlinear microscopy [24]. It comprises an infrared femtosecond laser oscillator which produces ~200 fs long pulses at a repetition rate of 50 MHz, generating an average power of 1 W. These pulses are tightly focused on to the bio-sample, using a high NA microscope objective, and a telescope system (lenses) to ensure overfilling of the entrance aperture of the objective. Typically, the bio-sample is irradiated with a maximum of 30 mW of incident laser power; it is scanned by using galvanometric mirrors and the two-photon induced fluorescence is collected in the backward direction via a lock-in amplifier that helps ensure high signal-to-noise ratio (S/N). The transmitted beam is directed into a monochromator and, thence, to a photo-detector. The monochromator makes it possible to monitor second-harmonic and third-harmonic signals in the forward direction.

Efficient use has been made of such combination nonlinear microscopy [24] to carry out in vivo imaging of cell morphology as well as various cellular processes in a nematode, *C. elangans*. A combination of two-photon induced fluorescence, second-harmonic, and third-harmonic microscopy has also enabled 3D image reconstruction to be carried out which permitted delineation of certain cell types, like neurons, from organs, like the pharynx, opening the way for quantitative studies to be performed on general problems concerned with neuronal degeneration.

3.5 Super-Resolution Microscopy: "Manipulation" of Molecular Spectra

Although modern technology—based on computer-aided design and fabrication (including grinding)—has revolutionized the manufacture of optical components, glass-based microscopes still face a fundamental barrier as far as the ultimate attainable resolution is concerned. This barrier is fundamental as it is a consequence of the wave nature of light, specifically relating to the diffraction of visible light wavefronts as they propagate through a microscope objective's circular aperture at the entrance focal plane. Two centuries ago, Abbé discovered that as light, of wavelength λ, propagates through a medium whose refractive index is n and focuses to a spot with angle θ, the radius, d of the diffraction limited focused spot will be [48, 49]

$$d = \frac{\lambda}{2n\sin(\theta)} = \frac{\lambda}{2[NA]}. \tag{3.4}$$

The term $n\sin(\theta)$ is the NA and, in contemporary optical systems, it can readily attain values in the vicinity of 1.5. Correspondingly, the Abbé limit can be written as $d = \frac{\lambda}{3}$.

In the case of green light (wavelength ~532 nm), a microscope objective whose NA is a modest 1 yields an Abbé limit of d ~260 nm. This is, of course, a small number in comparison to most cells (whose diameters would range from about 2 to ~100 μm). However, it is a large number when comparison is made to other biological entities, such as viruses (whose typical diameter may be of the order of 100 nm), proteins (which are even smaller in size, typically of the order of 10 nm), and a host of biomolecules of typical size 1 nm. All such bio-samples fall outside the purview of conventional microscopy, all because of the Abbé limit. Of course it is possible to increase the attainable resolution by making use of shorter wavelength light, as is done in UV and X-ray microscopes, but these are techniques that are not readily amenable to adoption in individual laboratories: they require large-scale facilities and they continue to suffer from disadvantages like expense, lack of contrast in biological samples, and the risk of photodamage.

In the last few years there has developed a new generation of microscopy that has sought to circumvent the Abbé diffraction limit. This so-called "super-resolution" microscopy relies on physical manipulation of molecular spectra. The diffraction limit along two orthogonal directions, $d_{x,y}$, quantifies the extent of blurring of a point source that is imaged through a lens. This blurring is usually expressed as the point spread function (PSF) [48]. The best attainable resolution in the axial direction for a confocal microscope can be estimated as [49, 50]

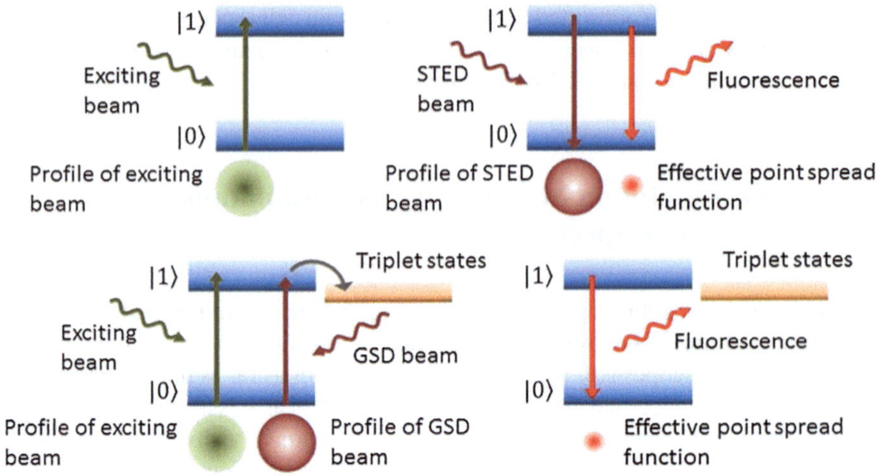

Fig. 3.4 *Upper panel* Stimulated emission depletion (STED) microscopy employs an additional red-shifted STED laser to pump molecules from excited "ON" states to ground "OFF" states before the molecule can emit a photon. The sample can be simultaneously scanned with excitation and STED lasers, and the resolution is determined by the peak power of the STED beam. *Lower panel* Ground state depletion (GSD) microscopy uses a single wavelength laser beam shaped as a doughnut, first at high intensity to "switch off" molecules by exciting them to triplet states, and then using the excitation laser to collect light from the remaining molecules (in the central zero of the doughnut). Note that the effective PSFs are impressively smaller in both cases

$$d_z \sim \frac{2n\lambda}{[n\sin(\theta)]^2} = \frac{2n\lambda}{[NA]^2}. \tag{3.5}$$

With green light propagating through a high NA ($\sim$1.5) objective, the resolution is limited to $\sim$150 nm and $\sim$350 nm in the lateral and axial directions, respectively. In a conventional microscope, the axial resolution is lowered further due to the geometrical optical effects. Super-resolution microscopy circumvents these limits by making use of the fact that the PSF depends only on the convolution of the excitation profile and the emission state of the molecules that are sought to be imaged [51, 52]. Figure 3.4 depicts the essential features of one variant of this type of microscopy that has come to prominence as stimulated emission depletion (STED) microscopy.

From an experimentalist's viewpoint, perhaps the easiest approach to sharpening the fluorescence focal spot in a microscope image is to selectively suppress that part of the fluorescence signal that lies on the outer edges of the focused spot. If such truncation can be attained on a focused spot that is otherwise diffraction-limited, it then becomes possible to circumvent the diffraction limit, d, since scanning with a smaller fluorescent spot automatically implies the attainment of enhanced spatial resolution. The method utilized to suppress fluorescence is stimulated emission, giving rise to the acronym STED (Fig. 3.4).

Experimentally, molecules which are in their ground electronic state are photoexcited using a conventional excitation laser beam (Fig. 3.4). A second, red-shifted laser beam, whose beam profile is made to be hollow in shape (a doughnut shape), is used to pump these electronically excited molecules into the ground state on ultrafast timescales, *before* there is time for the excited molecule to de-excite by emitting a photon. The spatial profile of the STED laser pulse—of doughnut shape—ensures that only those molecules that are located at the periphery of the focused spot are suppressed while, in the central portion of the doughnut beam, fluorescence remains unaffected. It is fluorescence only from this central portion that is made available for imaging with conventional instrumentation. The biomolecules that are to be imaged are normally scanned simultaneously by both laser beams, with the effective resolution being determined by the peak power of the STED beam. The full width at half maximum (FWHM) of the central portion, in which fluorescence is permitted, is expressed as [52]

$$d_{x,y} \sim \frac{\lambda}{2\eta \sin(\alpha)} \left[1 + \frac{a I_{\max}}{I_{se}}\right]^{\frac{1}{2}}, \tag{3.6}$$

where $I_{\max}$ is the peak intensity of the STED beam, I_{se} is the saturation intensity for stimulated emission, and a is a parameter that is related to the beam's intensity profile [52]. It is clear that the attainable effective resolution, d, scales as $\left(\frac{I_{\max}}{I_{se}}\right)^{\frac{1}{2}}$ while I_{se} itself scales as the inverse of the lifetime of the electronically excited states.

The choice of states to be targeted in STED microscopy opens many possibilities but it also presents some experimental constraints. Typically, stimulated emission is supposed to occur on nanosecond timescales and, consequently, values of I_{se} need to be higher ($\sim$100 MW cm^{-2}) than what microscopists are used to [50]. Such high values, of course, necessitate the use of femtosecond laser sources.

In an alternative strategy, known as ground state depletion (GSD) microscopy, biomolecules are electronically excited to triplet states whose lifetimes are much longer (typically, they lie in the ms and μs range). This leads to the possibility of using much lower values of I_{se} ($\leq$100 kW cm^{-2}) [53]. Both the STED and GSD processes are shown in cartoon form in Fig. 3.4. Excellent values of lateral resolution, typically $\sim$30 nm, are routinely obtainable using either of these methods; a value as low as 2.8 nm has also been reported [54].

On its own, neither STED nor its variants (like GSD) actually circumvent the Abbé diffraction limit because the laser beams utilized to accomplish the depletion are both diffraction-limited themselves. The physics that is required for effectively circumventing the Abbé diffraction limit lies in using stimulated emission to *saturate* the suppression of fluorescence. Increasing the intensity of the STED pulse—by using amplified femtosecond laser systems—helps accomplish this. Concomitantly it also ensures that the effective PSF (Fig. 3.4) can be made small enough (in diameter) to reach, in principle, the size of a single molecule!

The range of applications of STED microscopy and its variants is witnessing rapid and widespread increase in diverse areas of contemporary biophotonics [50], ranging from cell biology [55] to addressing issues concerning chromosome architecture [56], and applications in neurobiology, microbiology, and virology [57–59]. By way of example, we draw attention to the application of STED to obtain images of the distribution of protein molecules *within* cells in a two-colour experiment that enabled attainment of a spatial resolution of ~30 nm in the focal plane [60]. This remarkable value of resolution constitutes almost an order of magnitude improvement over typically obtained resolution with conventional confocal microscopy.

Two-colour STED microscopy was used to image cultured hippocampal neurons from neonatal rats [60] by staining with fluorescence dyes two synaptic proteins, Syntaxin 1 and Synaptophysin. The former is a protein found in the plasma membrane and is known to be involved in release of neurotransmitters. The latter is an important constituent of synaptic vesicles [61]. The images revealed, for the first time, the existence of protein clusters of ~41 nm size within the cell. The proteins were seen to apparently cluster in different locations within each cell, corresponding to locations in different structures of the synapse. The nanometre dimension spatial confinement of the synapse had precluded the application of conventional microscopy. Experiments that probe in three dimensions the co-localization of proteins within the interior of living cells were also extended to obtain hitherto-unavailable mapping of doubly-stained neurofilaments in differentiated neuroblastoma cells [60]. Neurofilaments constitute a major part of the axonal cytoskeleton; they comprise three subunits, designated neurofilament light, neurofilament medium, and neurofilament heavy. Mature filaments have incorporated within them the protein, α-internexin, and STED was successful in differentially imaging both α-internexin (using a red laser beam) and neurofilaments (using a green laser beam).

3.6 4D Microscopy and 4D Electron Microscopy

There has been a paradigm shift in the area of microscopic visualization of subcellular structures in the course of the last decade, with increasing use being made of fluorescing molecules infused within cellular structures. Detection of light from such fluorescing molecules—fluorophores—has made modern fluorescence microscopy a workhorse in many life sciences laboratories. The key elements of this type of microscopy are:

(i) illumination of the cell with light of appropriate wavelength that will be absorbed by the fluorophore molecules, and
(ii) detection of the resulting emission, at longer wavelengths, as the fluorophore molecules de-excite to a lower-energy state [62].

Enhancement of both the impact and the usefulness of such fluorescence-based microscopy is achieved when it is applied in so-called 4D mode. This enhancement is achieved by recording 3D images—which cover the three spatial dimensions—in time-dependent fashion. Microscopic images recorded as a function of time enable studies of the *dynamics* of cellular systems and processes. By measuring, in three dimensions, the fluorescence that is emitted from the irradiated sample, it becomes possible to readily quantify the spatial distribution of fluorophore molecules embedded within the sub-cellular structure that is being imaged. There have been two different approaches to accomplish such detection:

(i) scan the volume of interest within the cellular structure along the x, y, z coordinate axes or,
(ii) acquire 2D images—that are measured along the x, y axes—whilst a scan is made along the z (depth) axis.

Hence, 4D microscopy involves making measurements of a set of 3D images at discrete and regular time intervals. By employing a sufficiently short time interval between successive 3D measurements, it has been possible to investigate a variety of spatio-temporal dynamics, for instance, in meiosis [63], mitosis [64], endocytosis [65], cytogenesis [66], and in studying the breakdown and re-assembly of nuclear envelopes [67, 68]. However, this type of microscopy, dependent as it is on fluorescence, has, until very recently, failed to be applicable to those types of cells—such as yeast cells—that are intrinsically photosensitive [69]. The reason for such failure is of interest as it lies in the nature of the photoexcitation process itself.

Measurement of the intensity of fluorescence in such measurements depends on excitation dose, ζ:

$$\zeta = I\tau, \tag{3.7}$$

the product of excitation light intensity, I, and its duration, τ. The S/N ratio of the fluorescence signal scales with ζ, suggesting that the highest possible dose should be used. However, it is quite likely that large excitation doses will affect the irradiated cell's normal functioning through the toxic effect of fluorophore molecules [70]. Such toxicity is, in essence, a consequence of a photophysical process: it arises from the photon-induced (or, more likely, multiphoton-induced) dissociation of H_2O molecules which are omnipresent in live biological media. Such dissociation results in formation of O-atoms or OH radicals, both of which are, intrinsically, highly reactive. The magnitude of such phototoxicity is directly proportional to ζ, suggesting that the minimum possible dose should be used to ensure that the cell that is being microscopically probed continues to function in normal fashion (the cell continues to be "live" and not photo-damaged).

This constraint on the value of ζ has, until quite recently, resulted in a large class of cellular structures and processes remaining outside the ambit of 4D microscopy. However, a recent advance in deconvolution methodology [71] seems to have opened up the possibility of being able to retrieve microscopy images that are of sufficient resolution in the presence of a large amount of noise. Thus, it may now be possible to employ modest values of ζ—so as not to induce phototoxic effects in yeast (and other) cells—but, at the same time, still permit relatively sharp images to be retrieved by making feasible 3D measurements over long enough periods of time to enable proper 4D imaging.

It is a certainty that further developments in biophotonics, particularly in imaging technology, will be at the fore in determining how this field develops in the coming years. Already there are hints of several possible new avenues to be explored, such as using high quantum efficiency, electron multiplying CCD cameras with >90 % photon detection efficiency in conjunction with high NA optics that enable near unit collection efficiency for photons emitted by an irradiated cell. An additional benefit of technological advances in detectors, image capturing software, and fast data transfer is the possibility of directly imaging ultrafast processes.

A very recent variant of fluorescence-based 4D microscopy is 4D electron microscopy [72] whose development augers well for imaging of ultrafast phenomena in a variety of samples, including bio-samples. Electron microscopy of biological entities is conventionally carried out using phase-contrast imaging of specimens that are cryogenically cooled [73]. Over the decades electron microscopy has been truly embraced as a technique of very considerable utility in the life sciences but, at the same time, it is acknowledged to also impose strict constraints on (i) the coherence of the electron beam, (ii) the mandatory use of cryo-microscopy stages, (iii) the practical requirement of capturing a very large number (typically several thousands) of images in order to achieve adequately reliable reconstruction of the original structure of the biological entity, and (iv) no possibility of real-time imaging. There are several variants of 4D ultrafast electron microscopy which seek to overcome the first three of these constraints. They mostly depend on single electron imaging to visualize dynamics occurring on ultrafast timescales.

The technique essentially relies on using modestly intense femtosecond optical pulses, typically of 800 nm wavelength, to irradiate a photocathode which generate electron pulses; another optical pulse initiates a chemical change in the bio-sample by means of either a photoexcitation process or via a conventional T-jump. Figure 3.5a is a schematic depiction of an apparatus developed by Zewail and coworkers. Here the third-harmonic of the laser light is used to generate the ultrafast probe electron pulses at the photocathode whereas the second-harmonic beam is used to electronically excite the irradiated sample with a well-defined time delay with respect to the probe electron beam. The electron beam transmitted through the sample can be recorded either as an image, or an electron diffraction pattern, or, indeed, in the form of an electron kinetic energy spectrum.

The time delay between the electron packet and optical pulse that initiates photoexcitation determines the timescale of this type of microscope. The important point about this technique is that the temporal response of the detector that is used to acquire the signals is no longer relevant. The experimental arrangement attempts to carry out imaging using only a single electron at a time (only one electron is emitted when the photocathode is irradiated by the femtosecond laser pulse), so that space charge effects are of no consequence and, consequently, one does not have to worry about possible degradation of the coherence and imaging capability of such a microscope.

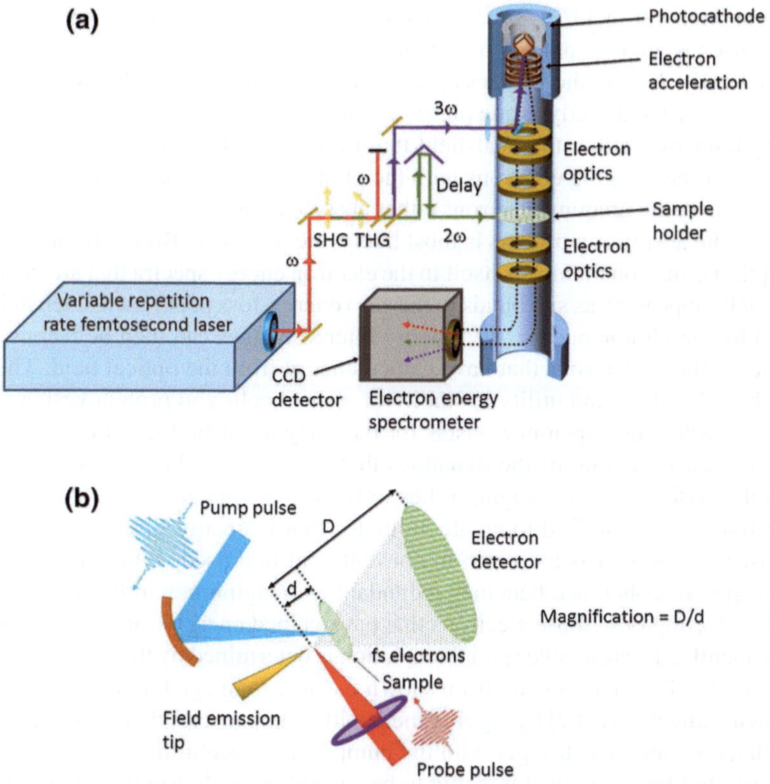

Fig. 3.5 **a** Schematic depiction of a 4D electron microscope. This is a variant of a conventional pump-probe technique and involves a femtosecond laser pulse producing an ultrafast pulsed electron beam upon irradiation of a photocathode. The kinetic energy of the electron pulse is probed as it transmits through the sample which is electronically excited by a pulse of second-harmonic light from the same femtosecond laser [74]. The sample can be as small as a few nanometres in size. Varying the time delay between the excitation laser pulse and the probe electron pulse and recording energy filtered images for each temporal delay yields the sample's electronic structure. **b** Another variant of the 4D electron microscope is the femtosecond photoelectron point projection microscope [75]. Here, the pump beam electronically excites the sample while the probe beam is used to generate the ultrafast electron pulses by excitation from a field emission tip. These electrons are electrostatically accelerated towards the sample. Upon transmission through the sample they are detected at a distance *D* from the source

Such single-electron imaging has now been successfully applied to most variants of electron microscopes, enabling the imaging of ultrafast dynamics in nanomaterials, with nanoscale resolution [76–78]. The addition of an electron energy spectrometer augments this type of microscope (Fig. 3.5a), giving rise to what has come to be known in the literature as femtosecond electron energy loss spectroscopy (FEELS) [79].

An application of FEELS that is likely to find increasing utility in the biological sciences is photon-induced near-field electron microscopy [72, 74]. Figure 3.5b schematically depicts the business part of the technique [75]: a femtosecond laser pulse is utilized to directly excite plasmons either in a nanoparticle or at an interface. During laser excitation, the near-field that develops at the interface permits direct coupling of the imaging electrons with the photons used to carry out the excitation. This results in the imaging electrons either absorbing or emitting integer numbers of photons (the absorption process is most likely due to inverse Bremsstrahlung). Such absorption/emission manifests itself in the electron energy spectra that are measured, with peaks appearing as sidebands to the zero energy-loss peak, each sideband being spaced by the photon energy, $h\nu$. Energy filtered images can then be reconstructed using only those electrons that have gained energy from the optical field. The technique has already found utility in studies of stained cells and protein vesicles [80].

These techniques open new vistas for the imaging of biological entities as well as for ultrafast imaging of the dynamics that occur in biophotonic structures. The principle advance is that imaging relies only on electrons that have absorbed photons from the optical field to create an image that is energy-filtered. Unlike phase contrast imaging that is the norm in conventional microscopy, here the need for a high degree of coherence becomes redundant as imaging now relies essentially on simply selecting only those electrons that have gained energy from the optical field. Consequently, the attainable spatial resolution is determined by the coherence length of the near-field. It has already been shown feasible to image biomolecules in aqueous environments [81, 82] by tagging them with nanoparticles that have been chosen such that they interact strongly with the pump light. Possibilities have already been demonstrated for dynamical imaging to be carried out of live cells in real time—on ultrafast timescales.

References

1. K. Xu, G.S. Zhong, X.W. Zhuang, Science **339**, 452 (2013)
2. T. Ishitsuka, Y.M. Li, R. Fischer, N. Takeshita, G.U. Nienhaus, Biophys. J. **104**, 652A (2013)
3. M. Coelho, N. Maghelli, I.M. Tolić-Norrelykke, Integr. Biol. **5**, 748 (2013)
4. W. Denk, J.H. Strickler, W.W. Webb, Science **248**, 73 (1990)
5. S. Kalies, K. Kuetemeyer, A. Heisterkamp, Biomed. Opt. Express **2**, 805 (2011)
6. K. König, P.T.C. So, W.W. Mantulin, B.J. Tromberg, E. Gratton, J. Microsc. **183**, 197 (1996)
7. A. Ustinoe, D.W. Piston, J. Microsc. **243**, 221 (2011)
8. M.E. Dickenson, E. Simbuerger, B. Zimmermann, C.W. Waters, S.E. Fraser, J. Biomed. Opt. **8**, 329 (2003)
9. C. Xu, W. Zipfel, J.B. Shear, R.M. Williams, W.W. Webb, Proc. Natl. Acad. Sci. USA **93**, 10763 (1996)
10. M. Drobizhev, S. Tillo, N.S. Makarov, T.E. Hughes, A. Rebane, J. Phys. Chem. B **113**, 855 (2009)
11. S.E. Tillo, T.E. Hughes, N.S. Makarov, A. Rebane, M. Drobizhev, BMC Biotechnol. **10**, 6 (2010)
12. M. Drobizhev, N.S. Makarov, S. Tillo, T.E. Hughes, A. Rebane, Nat. Methods **8**, 393 (2011)
13. X. Shu, A. Royant, M.Z. Lin, T.A. Aguilera, V. Lev-Ram et al., Science **324**, 804 (2009)

14. H.M. Kim, C. Jung, B.R. Kim, S.-Y. Jung, J.H. Hong et al., Angew. Chem. Int. Ed. **46**, 3460 (2007)
15. H.M. Kim, B.R. Kim, J.H. Hong, J.-S. Park, K.J. Lee, B.R. Cho, Angew. Chem. Int. Ed. **46**, 7445 (2007)
16. N. Soh, Anal. Bioanal. Chem. **386**, 532 (2006)
17. A. Gomes, E. Fernandes, J.L.F.C. Lima, J. Fluoresc. **16**, 119 (2006)
18. H. Cao, M.D. Heagy, J. Fluoresc. **14**, 569 (2004)
19. E.J. O'Neil, B.D. Smith, Coord. Chem. Rev. **250**, 3068 (2006)
20. E. Rebollo and C. Gonzalez, Curr. Protoc. Stem Cell Biol. **13**, 1H.2.1 (2010)
21. V. Kohli, K. Rehn, S. Sumanas, J. Vis. Exp. **56**, e3172 (2011)
22. N. Russek-Blum, H. Nabel-Rosen, G. Levkowitz, J. Vis. Exp. **46**, e1902 (2010)
23. L. Carvalho, C.P. Heisenberg, Methods Molec. Biol. **546**, 273 (2009)
24. G. Filippidis, E.J. Gualda, M. Mari, K. Troulinaki, C. Fotakis, N. Tavernarakis, Micron **40**, 876 (2009)
25. N. Ji, J.C. Magee, E. Betzig, Nat. Methods **5**, 197 (2008)
26. C.J. Engelbrecht, R.S. Johnston, E.J. Seibel, F. Helmchen, Opt. Express **16**, 5556 (2008)
27. B.A. Flusberg, E.D. Cocker, W. Piyawattanametha, J.C. Jung, E.L.M. Cheung, M.J. Schnitzer, Nat. Methods **2**, 941 (2005)
28. A.M. Lee, H. Wang, Y. Yu, S. Tang, J. Zhao et al., Opt. Letters **36**, 2865 (2011)
29. A. Singh, S. Kumar, V.P. Singh, J. Balaji, Cur. Sci. **105**, 1537 (2013)
30. J.D. Sipe, *Amyloid Protein: The Beta Sheet Conformation and Disease* (Wiley and Sons, New York, 2005)
31. P. Hanczyc, M. Samoc, B. Norden, Nature Photon. **10**, 1038 (2013)
32. E. Collini, Phys. Chem. Chem. Phys. **14**, 3725 (2012)
33. J.F. Smith, T.P.J. Knowles, C.M. Dobson, C. MacPhee, M.E. Welland, Proc. Natl. Acad. Sci. USA **103**, 9196 (2006)
34. H.A. Lashuel, P.T. Lansbury, *Are Amyloid Diseases Caused by Protein Aggregates that Mimic Bacterial Pore-forming Toxins?* (Cambridge University Press, Cambridge, 2006)
35. S.R. Collins, A. Douglass, R.D. Vale, J.S. Weissman, PLoS Biol. **2**, 1582 (2004)
36. G. Bitan, A. Lomakin, D.B. Teplow, J. Biol. Chem. **276**, 35176 (2001)
37. J. Dong, R. Revilla-Sanchez, S. Moss, P.G. Haydon, Neuropharmacology **59**, 268 (2010)
38. W.E. Klunk, Neurobiol. Aging **19**, 145 (1998)
39. B.J. Bacskai, G.A. Hickey, J. Skoch, S.T. Kajdasz, Y. Wang et al., Proc. Natl. Acad. Sci. USA **100**, 12462 (2003)
40. C.A. Mathis, Y. Wang, D.P. Holt, G.-F. Huang, M.L. Debnath, W.E. Klunk, J. Med. Chem. **46**, 2740 (2003)
41. L. Moreaux, O. Sandre, J. Mertz, J. Opt. Soc. Am. B **17**, 1685 (2007)
42. A. Volkmer, J.X. Cheng, X.S. Xie, Phys. Rev. Lett. **87**, 023901 (2001)
43. S. Fine, W.P. Hansen, Appl. Opt. **10**, 2350 (1971)
44. E. Brown, T. McKee, E. DiTomaso, A. Pluen, B. Seed et al., Nat. Methods **9**, 796 (2003)
45. J. Mertz, L. Moreaux, Opt. Commun. **196**, 325 (2001)
46. R.M. Williams, W.R. Zipfel, W.W. Webb, Biophys. J. **88**, 1377 (2005)
47. D.R. Larson, W.R. Zipfel, R.M. Williams, S.W. Clark, M.P. Bruchez et al., Science **300**, 1434 (2003)
48. E. Abbe, J. Roy. Microscop. Soc. **3**, 790 (1883)
49. L. Novotny, B. Hecht, *Principles of Nano-optics* (Cambridge University Press, Cambridge, 2006)
50. R. Roy, Cur. Sci. **105**, 1524 (2013)
51. S.W. Hell, Science **316**, 1153 (2007)
52. S.W. Hell, Nat. Methods **6**, 24 (2009)
53. S. Bretschneider, C. Eggeling, S.W. Hell, Phys. Rev. Lett. **98**, 218103 (2007)
54. D. Wildanger, B.R. Patton, H. Schill, L. Marseglia, J.P. Hadden, et al., Adv. Mater. **24**, OP309 (2012)
55. B. Huang, M. Bates, X. Zhuang, Ann. Rev. Biochem. **78**, 993 (2009)

56. C. Flors, W.C. Earnshaw, Curr. Opin. Chem. Biol. **15**, 838 (2011)
57. J. Tonnesen, U.V. Nagerl, Exp. Neurol. **242**, 33 (2013)
58. D.I. Cattoni, J.B. Fiche, M. Nollmann, Curr. Opin. Microbiol. **15**, 758 (2012)
59. C. Coltharp, J. Xiao, Cell Microbiol. **14**, 1808 (2012)
60. L. Meyer, D. Wildanger, R. Medda, A. Punge, S.O. Rizzoli et al., Small **4**, 1095 (2008)
61. T. C. Südhof, Annu. Rev. Neurosci. **27**, 509 (2004)
62. S. Prasad, R. Swaminathan, Cur. Sci. **105**, 1549 (2013)
63. J. Fung, W. Marshall, A. Dernburg, D. Agard, J. Sedat, J. Cell. Biol. **141**, 5 (1998)
64. C. Rieder, A. Khodjakov, Science **4**, 91 (2003)
65. S. Kumari, S. Mayor, Nat. Cell Biol. **10**, 30 (2008)
66. R. Pelham Jr., F. Chang, Nature **419**, 82 (2002)
67. J. Beaudouin, D. Gerlich, N. Daigle, R. Eils, J. Ellenberg, Cell **108**, 83 (2002)
68. R. Moira, M. Yoona, S. Khuona, R. Goldman, J. Cell. Bol. **151**, 1155 (2000)
69. M. Arigovindan, Cur. Sci. **105**, 1501 (2013)
70. D.J. Stephens, V.J. Allen, Science **300**, 82 (2003)
71. M. Arigovindan, J.C. Fung, D. Elnatan, V. Mennella, Y.-H.M. Chan et al., Proc. Natl. Acad. Sci. USA **10**, 17344 (2013)
72. B. Barwick, A.H. Zewail, ACS Photonics **2**, 1391 (2015)
73. R.M. Glaeser, R.J. Hall, Biophys. J. **100**, 2331 (2011)
74. F. Carbone, B. Barwick, O.-H. Kwon, H.S. Park, J.S. Baskin, E.H. Zewail, Chem. Phys. Lett. **468**, 107 (2009)
75. E. Quinonez, J. Handali, B. Barwick, Rev. Sci. Instrum. **84**, 103710 (2013)
76. H.S. Park, J.S. Baskin, O.-H. Kwon, A.H. Zewail, Nano Lett. **7**, 2545 (2007)
77. M. Gulde, S. Schweda, G. Storeck, M. Maiti, H.K. Yu et al., Science **345**, 200 (2014)
78. M. Müller, A. Paarmann, R. Ernstorfer, Nature Commun. **5**, 5292 (2014)
79. F. Carbone, O.-H. Kwon, A.H. Zewail, Science **325**, 181 (2009)
80. D.J. Flannigan, B. Barwick, A.H. Zewail, Proc. Natl. Acad. Sci. USA **107**, 9933 (2010)
81. D.R. Glenn, H. Zhang, N. Kasthuri, R. Schalek, P.K. Lo et al., Sci. Rep. **2**, 865 (2012)
82. N. de Jonge, D.B. Peckys, G.J. Kremers, D.W. Piston, Proc. Natl. Acad. Sci. USA **106**, 2159 (2009)

Chapter 4
Ultrafast Single-Molecule Spectroscopy

Abstract The detection and spectroscopic studies of individual molecules has begun to find widespread applications in the life sciences. While conventional time-resolved experiments continue to provide useful information about ensemble-averaged properties, ultrafast single molecule techniques are able to track the photodynamics of *single* molecules, revealing their unique transient intermediates. This chapter presents an overview of single molecule techniques like surface enhanced coherent anti-Stokes Raman scattering, Förster resonance energy transfer, pump-probe spectroscopy, and pulse-shaping. These techniques can not only identify but also provide information about electronic/vibrational wavepacket interferences and relaxation mechanisms at the level of individual molecules. As an illustration of their relevance in ultrafast biophotonics, a study of persistence of coherence in an individual photosynthetic complex under physiological conditions is presented.

4.1 Introduction

Many single-molecule techniques, such as mass spectrometry [1], have been around for five decades or longer. But these usually work only when samples are in the gas phase. On the other hand, detection of single molecules in the condensed phase has become practicable relatively recently [2–4]. In complex and heterogeneous condensed-phase systems of the type invariably encountered in the life sciences, the optical response of the constituent molecules is strongly influenced by the local environment. This results in variation of molecular properties like resonance wavelength, dipole moment, polarization, fluorescence lifetimes and reaction products [5–10]. Accordingly, several steady-state experiments have consistently shown that chemically identical molecules can exhibit environment-dependent varying optical responses. Thus, the major challenge in biophotonic investigations of biomolecules is to either ensure identical surrounding for all molecules comprising the bio-system under study, or to discriminate a weak signal from a *single* molecule against a large background. Monitoring individual molecules independently (i) eliminates ensemble averaging, (ii) enables determination of intrinsic properties, and (iii) reveals the effects of local environment [9, 10]. Furthermore, following the dynamics of

P. Vasa and D. Mathur, *Ultrafast Biophotonics*, Biological and Medical Physics, Biomedical Engineering, DOI 10.1007/978-3-319-39614-9_4

individual molecules obviates the need to synchronize the entire complex system and provides insights into environment-dependent intermediate states in an individual photocycle [11].

In the case of steady-state single-molecule techniques, detecting the Stokes shifted fluorescence or Raman scattering signal arising from an individual molecule has proven to be a reasonably straightforward way to enable background-free identification [9, 10, 12]. These techniques have developed into being capable of molecular specificity, contrast and compatibility with multi-colour and live-cell imaging [13–17]. Moreover, these techniques have benefited from spectacular breakthroughs in nonlinear and super-resolution microscopy techniques [18, 19], discussed in Chap. 3, that now allow sub-wavelength scale imaging, down to ~50 nm, at video rates. Since Raman scattering cross-sections are extremely weak, an enhancement scheme like surface enhanced Raman scattering (SERS) is generally employed in conjunction with confocal microscopy to achieve sufficiently high signal-to-noise ratio [12]. Although fluorescence microscopy can successfully detect and identify single molecules, it is not suitable for fast dynamical studies [9, 10, 16, 17] as the highest achievable temporal resolution is limited by the spontaneous emission lifetime of the molecule or the attached fluorophore [9, 10]. Unfortunately, for most biomolecules and fluorophores, spontaneous emission is a relatively slow process, occurring on picosecond to nanosecond timescale. In contrast, biological processes like energy/charge transfer, electronic dephasing, isomerization, and vibrational dynamics all occur on femtosecond to picosecond timescales, much faster than the spontaneous fluorescence emission. As a result investigating ultrafast processes at the single-molecule level is beyond the capabilities of conventional fluorescence and Raman nanospectroscopy [9, 10].

Until recently, ultrafast processes were almost exclusively studied using spatially integrated, pump-probe [20–25] or multi-dimensional spectroscopy [26–29] schemes discussed in Chap. 7, which yield an ensemble-averaged response. Originally developed for investigating artificial and/or ordered samples, the implicit assumption underlying these spatially integrated schemes is that all molecules involved are identical and have no (or identical) interaction with the surrounding [9, 10]. While this assumption may be valid for many artificial structures, it is certainly not valid for most biological systems, which strongly interact with the environment. The central common feature of biological functionalities, like energy/charge transfer in photosynthesis and vision, is an arrangement of closely packed conjugated molecules with a high degree of conformational and electronic disorder [30, 31]. Therefore, it is very relevant to investigate ultrafast dynamics of such heterogeneous assemblies using techniques that allow one to differentiate the response of each individual molecule in its unique local environment. This chapter presents an overview of some spatially-resolved, ultrafast excitation schemes for analysis of individual quantum systems (like molecules, or moieties). As we shall see, it is becoming possible to investigate various aspects of molecular dynamics—like energy transfer, electronic coherence and wavepacket interference—using these single-molecule schemes.

4.2 Surface Enhanced Coherent Anti-Stokes Raman Scattering (SECARS)

Raman spectroscopy is a powerful tool that provides information about molecular structure via the measurement of vibrational modes in single molecules. It accesses processes that are fast (femtosecond to picosecond) and is molecule specific. However, Raman scattering cross-sections are extremely weak, $\sim 10^{12}$ times smaller than those for fluorescence [12, 32]. Therefore, unless an enhancement scheme is employed, detection of single molecules is close to impossible. Surface enhanced Raman scattering (SERS) has advanced Raman spectroscopy to the detection limit of individual molecules under optimized conditions [33–37]. SERS relies on field enhancement created by surface plasmon polaritons (SPPs) that are excited on metal nanostructures (Chap. 3) and weak emitter-SPP coupling [25, 38–41]. Existing theories of SERS are based on local field enhancement (in close proximity of metal nanostructures) and charge transfer complexes but the exact mechanism is still a matter of investigation [12, 32]. SERS is driven by the dual benefits of plasmonic structures:

(i) the SPP field enhancement magnifies the intensity of incident light and ensures better coupling with the Raman scatterer, and
(ii) there is further amplification by the improved out-coupling mechanism of the plasmonic structure, yielding an enhanced Raman signal.

Raman enhancements by factors as large as 10^{10}–10^{11} have been reported [12, 32]. The disadvantage is that an optimized nanostructure possessing a plasmon resonance that overlaps the excitation wavelength is mandatory, and enhancement is experienced by only those molecules that are in close proximity to the metal nanostructure. Also, it is very sensitive to light polarization and molecular orientation. Despite these limitations, SERS is now an established single-molecule technique, and a wide variety of nanostructured metal substrates supporting SPPs have been proposed or are commercially available [12]. To achieve high spatial resolution SERS is generally used in conjunction with a confocal microscope. To further improve the spatial resolution and signal strengths of SERS, a metal tip of an atomic force microscope has also been successfully employed as a plasmonic resonator [33–37]. The small tip radius ($\sim$10 nm) and the possibility of controlling the tip-molecule separation on nanometre lengthscales has enabled tip enhanced Raman scattering (TERS) to even map a single molecule [37]. Although TERS can provide much higher spatial resolution compared to a confocal microscope its depth of field is extremely narrow, limiting it to surface studies. For biological samples in natural surroundings, a confocal microscope with a wider depth of field and underwater capability permits investigations at even $\sim$100 μm depths; this often proves to be more advantageous compared to the superior resolution of scanning probe techniques.

SERS schemes are based on spontaneous Raman scattering, as depicted in Fig. 4.1a. It is possible to improve single-molecular detection sensitivity by combining plasmonic (SERS) enhancements with coherence in a stimulated Raman

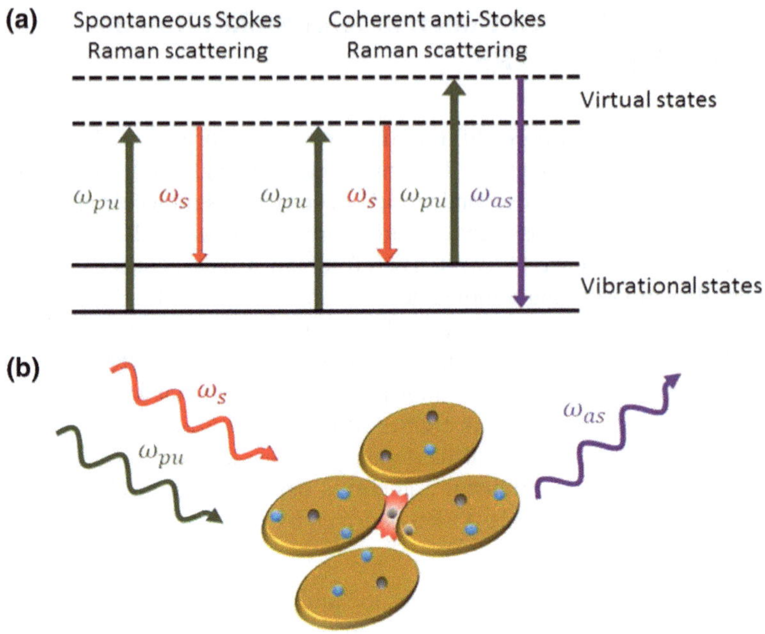

Fig. 4.1 **a** Schematic energy level diagrams depicting spontaneous Raman scattering and coherent anti-Stokes Raman scattering (CARS). **b** A surface enhanced CARS (SECARS) employing a gold nano-quadrumer comprising four gold discs of diameter ~160 nm. The ultrashort pump (ω_{pu}) and Stokes (ω_s) pulses generate a SECARS signal (ω_{as}) from a molecule in the central gap of the quadrumer [32]

scattering process. Coherent anti-Stokes Raman scattering (CARS) is one such third-order nonlinear optical process governed by molecular coherence [32, 42, 43]. In CARS, the pump and Stokes fields, with frequencies ω_{pu} and ω_s, interact coherently through stimulated Raman scattering that is governed by the third-order polarizability of a molecule, thereby generating an anti-Stokes signal ω_{as} with intensity, $I_{CARS} \propto I_{pu}^2 I_s$. Being a nonlinear process, it needs relatively high excitation intensities compared to spontaneous Raman scattering and, hence, CARS experiments are usually performed using ultrashort laser pulses. The intensity of the CARS signal is resonantly enhanced when $\omega_{pu} - \omega_s$ matches a Raman-active vibrational level in the molecule. This two-pulse, resonant as well as coherent process, schematically shown in Fig. 4.1a, is typically orders of magnitude stronger than spontaneous Raman scattering [42]. The CARS cross-sections can be further boosted by employing an appropriately designed resonant plasmonic structure [32, 43]. If one or more of ω_{pu}, ω_s and ω_{as} are in resonance with the collective modes of the plasmonic nanostructure, the surface enhanced CARS (SECARS) signal from molecules adsorbed onto the nanostructure experience the dual benefits of the local fields of the excited plasmon modes.

It has been shown that plasmonic nanostructures supporting Fano resonances can be designed to perform SECARS with single-molecule sensitivity [32]. The nanostructure used is a gold quadrumer—four coupled gold nanodisks—as shown in Fig. 4.1b, exhibiting a strong plasmonic Fano resonance [44–47]. The local fields in a single junction at the centre of the quadrumer structure give rise to highly localized SECARS enhancement—by a factor of $\sim 10^{11}$ relative to spontaneous Raman scattering! The single-molecule sensitivity of the technique has been verified using a statistically rigorous bi-analyte method, where para-mercaptoaniline (p-MA) and adenine molecules (both with Raman cross-sections of $\sim 10^{-30}$ cm^2 sr^{-1} in air for near-IR excitation) were clearly distinguishable from an extremely dilute mixture such that, on average, only one molecule was present in the quadrumer gap region [32]. The measurements on several quadrumer substrates showed that spectra were dominated by either one analyte or the other, or no molecules were detected at all. This confirmed that SECARS is observed only for a molecule present at the centre of the quadrumer, where the field enhancement is maximum [32].

Even though a confocal microscope can provide resolution of $\sim$500 nm, the strongly localized character of SECARS helps in substantially enhancing the spatial resolution, to $\sim$50 nm, and lowering the background as the molecules located just outside the gap do not contribute to the detected signal. As SECARS employs ultrashort laser pulses for excitation, it can be easily extended to perform time-resolved studies on ultrashort timescales. Though advantages and applications of SERS and SECARS have now been well established for single-molecule identification, microscopic understanding and theoretical modeling are yet to be developed. Also, fabrication of high quality plasmonic resonators relies on advanced lithography techniques capable of providing sub-10 nm resolution. Therefore, these techniques remain relatively expensive and difficult to implement at present. However, the high sensitivity and versatility offered by SECARS can undoubtedly lead to the identification and dynamical studies of unknown molecules at the ultimate limits of chemical sensitivity.

4.3 Förster Resonant Energy Transfer (FRET)

Within the past two decades, single-molecule fluorescence techniques have proven their potential, and are now routinely used in many biological investigations [48]. They provide data of direct relevance to statistical physics since information is now obtainable about the distributions and heterogeneity of structures in an ensemble, and their dynamics [10, 16, 17]. Two main types of single-molecule experiments are widely employed in biology: force-probe methods using atomic force microscopy (AFM) or optical tweezers [49], and fluorescence detection [50], especially in combination with Förster resonance energy transfer (FRET). AFM and optical tweezers have both proved to be remarkably versatile in probing the mechanical stability and folding dynamics of proteins. Here, we discuss single-molecule fluorescence

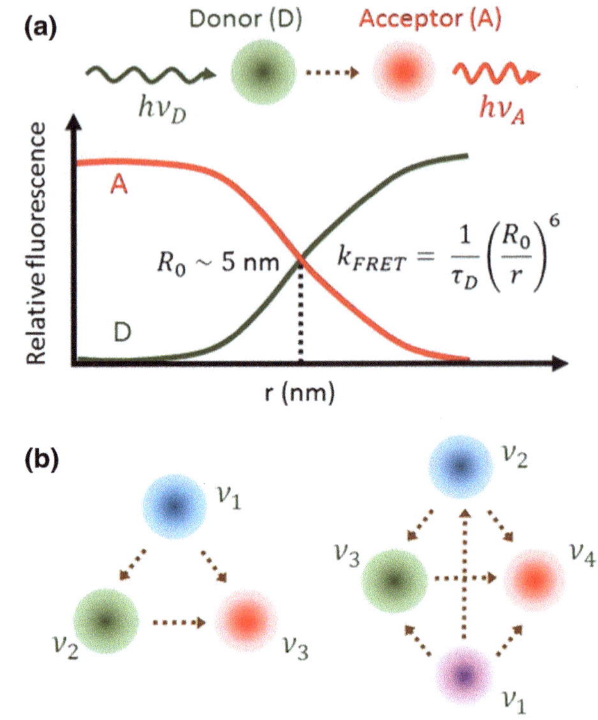

Fig. 4.2 Schematics of **a** single-colour Förster resonant energy transfer (FRET) with $\nu_D > \nu_A$ and **b** multi-colour FRET with $\nu_1 > \nu_2 > \nu_3 > \nu_4$. The latter permits simultaneous tracking of more than one dynamical event on nanometre length scales

spectroscopy with emphasis on FRET. The scope of FRET continues to expand, both in terms of questions addressed and in terms of accessible timescales [16, 17].

In its simplified classical description, FRET, schematically shown in Fig. 4.2a, is governed by the resonance of the transition dipole moments of donor (D) and acceptor (A) fluorophores. The rate constant of the D→A energy transfer (k_{FRET}) depends on (i) r^{-6}, with r being the D–A distance, (ii) the fluorescence lifetime of the donor (τ_D), and (iii) a characteristic distance R_0 that is deduced from the spectroscopic properties of D and A. Experimentally, the acceptor fluorescence is detected by a confocal microscope. Time-dependent changes in the acceptor fluophore signal provide information about the D–A distance [16, 17]. Due to the r^{-6} power dependence, FRET signals are extremely sensitive to displacement and rotation on the nanoscale. However, changes in signal manifest themselves only for relatively small separations (force) and rotations (torque), thus limiting the dynamic range of this technique. Typical values of R_0 (2–10 nm) make single-molecule FRET suitable for obtaining nanoscale distance information. A similar but complementary technique, relying on luminescence quenching, is called photo-induced electron transfer (PET). Here, static quenching of a fluorophore, typically by tryptophan, is used to obtain dynamics information based on contact formation within proteins or peptides. Unlike in FRET, the characteristic distances for PET are on sub-nanometre scales [16, 17].

One of the key challenges of contemporary single-molecule fluorescence imaging relates to the growing need to correlate multiple events in space and time. The function of most complex biological systems most often entails time-dependent changes in conformation and composition. If the goal is to understand the macromolecular mechanisms involved in all of its complexity, the observation of only one molecular species or the interaction of only a pair of species at a time, provides only a partial view and is insufficient [16, 17]. Fortunately, fluorescence techniques readily offer the possibility to simultaneously observe multiple processes through the use of spectrally distinct fluorophores [51, 52]. Owing to instabilities and photo-bleaching of the available fluorophores, however, their selection is often the performance-limiting factor. In this context, multi-colour single-molecule FRET (Fig. 4.2b) has recently begun to find its stride as a viable tool in biological investigations [53–55]. Using this technique, it is possible to simultaneously track translational and/or rotational dynamics of two or more domains. As a result of sustained efforts in dye and detector development, labeling chemistry, data analysis, and instrumentation design, multi-colour single-molecule FRET imaging has emerged as a unique technique to elucidate correlated conformational events in vitro and in living cells [16, 17, 56, 57].

Though versatile and capable of providing information on nanometre length scales, FRET is, intrinsically, not a time-resolved technique. However, dynamical information may be obtained using FRET by adding a time-resolved component like fluorescence correlation spectroscopy (FCS). FCS is based on measuring the fluctuations of fluorescence intensity in time followed by its correlation analysis [58]. FRET-FCS yields dynamical information on nanosecond to microsecond timescales and can be readily implemented in confocal instruments [16, 17, 58]. In FRET-FCS or PET-FCS experiments, the long-range distance fluctuations between the donor and the acceptor fluorophore (or quencher) provide information on the characteristic timescale of inter-conversion between different configurations of proteins or polypeptide chains [59–62]. Such combined methods can span more than fifteen orders of magnitude in time, from nanosecond to hours, allowing a wide range of biologically relevant processes—like protein folding—to be investigated. Quantitative FRET-FCS studies of unfolded and disordered proteins on sub-microsecond timescale have enabled rationalization of the folding mechanism in terms of polymer-physical concepts [59–62].

Time-correlated single photon counting (TCSPC) is another such technique which can be coupled with FRET to widen its scope and timescale [16, 52], particularly on the faster side. TCSPC is used to analyze the relaxation of molecules from an excited to a lower energy state [63, 64] which occurs via photon emission (fluorescence) in statistical fashion. By observing how long individual molecules take to emit photons, and then combining all these data points in an intensity versus time graph yields an exponential decay curve; typically, several individual excitation-relaxation events are recorded and then averaged to generate the curve [63, 64]. The assumption is that the relaxation mechanism is the same for every event and there is no photobleaching or blinking. A major complicating factor in TCSPC is that decay processes often involve multiple energy states, and thus multiple relaxation times. Even more complicating is the presence of inter-system crossing and other non-radiative processes in a molecule. Though non-linear least squared analysis and multi-exponential fits can usually yield the different time constants, the analysis can be complicated and is often susceptible to errors. Another drawback is that it is limited to studying energy states that result in fluorescent decay, whereas non-radiative relaxation pathways are often also important. TCSPC can provide time resolution of several picoseconds, limited only by the bandwidth of the photon-counting and correlating electronics. For higher time resolution, nonlinear techniques like fluorescence up-conversion [20–22], are needed but these are difficult to implement at the single-molecule level due to very weak signals.

4.4 Near-Field Ultrafast Pump-Probe Spectroscopy

A wide variety of elementary biological processes involve ultrafast molecular dynamics [30, 31, 65–67] that occur on sub-wavelength length scales. This makes direct space- and time-resolved imaging of these phenomena experimentally challenging. Spatially integrated ultrafast pump-probe spectroscopy techniques, discussed in Chap. 7, can provide time resolution in the femtosecond range [20–25, 68]. The main principle here is to use a short pump pulse to excite a set of molecules to a particular state, thus synchronizing a subset of the ensemble. Next, a second, generally weak and delayed, pulse probes the evolution of the created population. In most experiments, a fast optical response process, such as saturation of absorption or excited state absorption, is measured to track the system's evolution. A particularly powerful variant is 2D-spectroscopy, which can provide insight into coherent coupling between two electronic states [26–29]. In spite of the high temporal resolution, spatial resolution—even in case of confocal schemes—is inherently diffraction-limited by the wavelength of light and rarely goes beyond a few hundred nanometres. Electron- and X-ray-based microscopy techniques (Chaps. 3, 11) on the other hand, achieve much higher spatial resolution, down to the angstrom range [69–75], but are more instrumentation intensive. Space-time dynamics of electronic and vibrational wavepackets in molecules and nanostructures can thus be obtained only indirectly, often requiring additional *a-priori* knowledge about potential energy surfaces and the energetics of biomolecules.

Along with confocal microscopy, the last decade has seen development of near-field optical spectroscopy techniques [76–78] that are based on evanescent rather than propagating electromagnetic fields. Sub-wavelength optical resolution over the range 10–50 nm is now readily reached using aperture-based [79, 80] and scattering-type "apertureless" [81, 82] near-field probes. Combining near-field optical spectroscopy with ultrafast laser pulses thus promises the desired simultaneous high spatio-temporal resolution. Experimentally, the challenge in combining near-field scanning optical microscopy (NSOM), also known as scanning near-field microscopy or SNOM) with ultrafast laser spectroscopy consists in generating ultrashort bursts of light with sufficient temporal and sub-wavelength spatial resolution. Conceptually, the most straightforward solution to this problem is to funnel ultrafast light pulses through aperture-type near-field fibre probes [83, 84]. A reflective metal coating on the fibre tip aids better light throughput and spatial resolution. In most experiments, however, it turns out that the finite dispersion due to the fibre taper, rather than the pulse broadening induced by the aperture, limits the achieved time resolution. An alternative way to localize ultrafast laser pulses to spot sizes down to few nanometres is to make use of optical field enhancement at the apex of ultrasharp metal tapers [85–87]. This configuration is known as scattering-NSOM or s-NSOM (or s-SNOM). The optical resonances supported by metal tips typically lie in the visible to near-IR range and extend over more than 100 nm [76, 88], carrying sufficient bandwidth to support the generation of localized, sub-10 fs pulses. Depending upon requirements, different cw or pulsed laser sources can be used in near-field experiments to obtain dynamical information over a wide range of timescales.

A set-up of a typical ultrafast near-field pump-probe spectrometer is shown in Fig. 4.3. Laser pulses are sent through a pulse compressor for spectral shaping and pre-compensation of group velocity dispersion experienced in the near-field fibre probes before splitting into pump and probe pulses [76]. The pulse pair is then coupled into a near-field probe. The probe tip is maintained in close proximity of the sample (mounted on a x–y scanner) using the feedback arrangement of an AFM. The near-field light reflected from the sample is collected through the same fibre and coupled into a spectrometer or a detector. Several variants of the technique based on the design

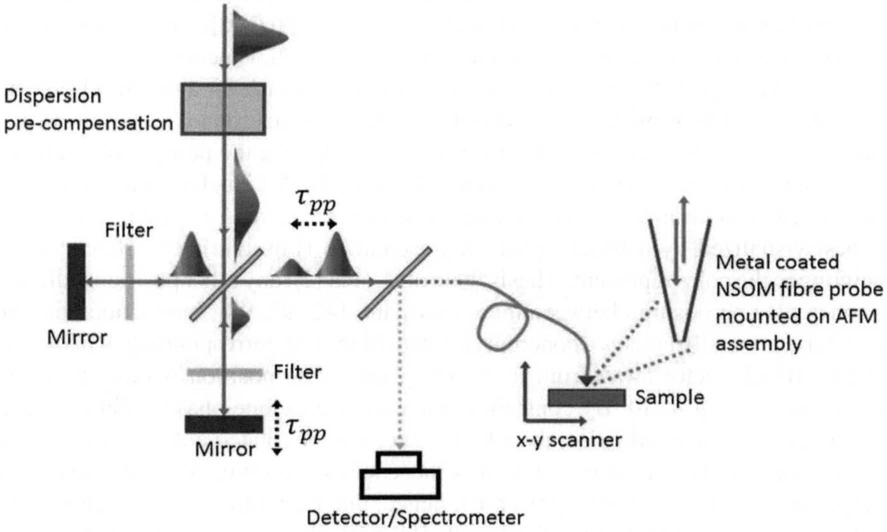

Fig. 4.3 Schematic of an ultrafast near-field scanning optical microscope (NSOM). Pump and probe pulses derived from an ultrafast laser are coupled into a metal coated near-field fibre probe after appropriate dispersion pre-compensation. The near-field reflected light is collected through the same fibre and coupled into a spectrometer. The fibre probe is maintained close to the sample which is mounted on a scanner using an AFM assembly. Different filters can be inserted for selecting pump and probe spectra

of the near-field probe and excitation/collection schemes have been proposed and implemented [76]. Apart from transient absorption measurements, this technique can also be extended to perform SERS [33–37], THz frequency comb Fourier transform spectroscopy [89, 90], and spatially resolved electron microscopy [91, 92].

While studies of material properties in semiconductor, organic photovoltaic heterostructures and conjugated polymers using near-field transient absorption microscopy are well established [93–95], the adoption of the technique to the life sciences is challenging due to constraints associated with aqueous environments, very limited depth of field (essentially meant for surface studies) and toxicity. Nonetheless, under appropriate experimental conditions, ultrafast near-field techniques are well suited for biological studies and pharmacological research, for example to visualize the spatial distribution of specific (macro)molecules in tissue or to investigate the spatial-temporal drug distribution and drug-uptake into cells [95].

4.5 Control of Single Molecule Electronic Coherence and Vibrational Wavepacket Interference

Despite the high potential of single-molecule or spatio-temporally resolved pump-probe spectroscopy, the lack of a phase relation between pump and probe pulses (particularly for non-degenerate pump and probe spectra) prevents the disentangling of contributions from coherence and dephasing effects [10, 96]. Clearly, a coherent excitation scheme, as in case of 2D-spectroscopy [26–29], is necessary.

Figure 4.4 depicts the basic notion of a coherent approach for a two-level system. A single molecule is resonantly excited from the electronic ground ($|1\rangle$) to an excited state ($|2\rangle$) by femtosecond double-pulse sequences. Unlike the pump-probe scheme discussed above, now the pulses are phase-locked [10, 96]. Such excitation induces stimulated absorption and emission and prepares the system in a certain state which is best visualized by a Bloch sphere representation (Fig. 4.4a). The Bloch vector pointing to the poles represents a basis-state of $|1\rangle$ and $|2\rangle$; any other position indicates a coherent superposition between these two states [42, 97, 98]. Interaction with the first pulse generally creates a coherent superposition state corresponding to a rotation of the Bloch vector away from its initial ground-state position without reducing its amplitude (Fig. 4.4b). By controlling the carrier envelope phase difference, $\Delta\varphi$ (Chap. 2), of the second pulse, the Bloch vector can be rotated either away from or towards the ground state (Fig. 4.4c). In the absence of dephasing, the state created by the pump pulse is indefinitely preserved, hence, the delay time τ_{pp} between the two pulses is unimportant [10, 96]. The length of the Bloch vector always remains unity. Now, if the molecule is surrounded by a disordered environment or in the presence of dephasing, the phase memory between the ground- and excited-state wavefunctions (or the coherent superposition state) is rapidly erased. As a result the magnitude of the Bloch vector reduces with time (Fig. 4.4d). Now, the second pulse can affect the Bloch vector rotation only if τ_{pp} is less than the characteristic dephsing time of the

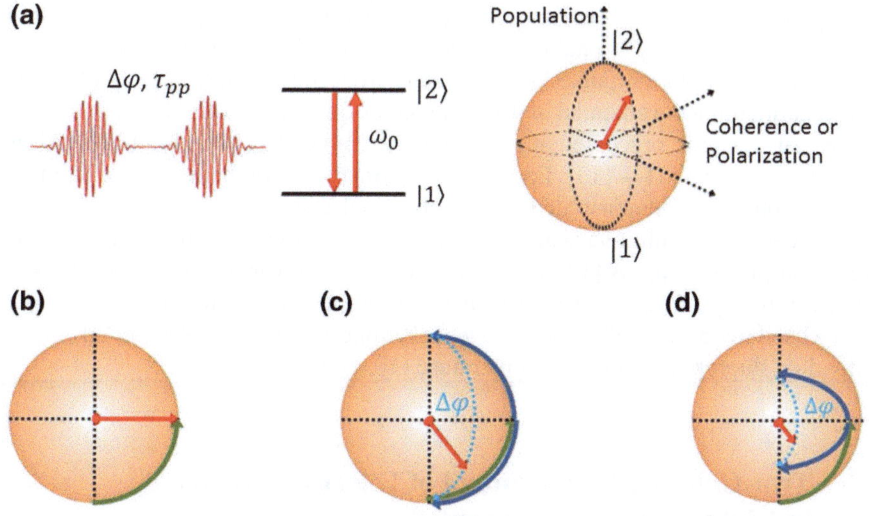

Fig. 4.4 **a** Schematic of a pure two-level system ($|1\rangle$ and $|2\rangle$) excited by a phase locked pulse pair with phase difference $\Delta\varphi$ and delay τ_{pp} and its Bloch sphere representation. **b** The first pulse creates a coherent superposition of the state corresponding to the rotation of the Bloch vector on the sphere. **c** In the absence of any disorder, the second pulse either increases the rotation or reduces it, without altering its length. The delay has no effect on the rotation. **d** In the presence of disorder, the phase memory is rapidly erased, resulting in reduction of the length of the Bloch vector. Now, both $\Delta\varphi$ and τ_{pp} affect its length and rotation. If the delay is longer than the dephasing time of the system, no phase memory is preserved and the second pulse cannot induce coherent rotation

system. Therefore, in the presence of disorder/dephasing the final position of the Bloch vector tip after the pulse sequence becomes a function of both $\Delta\varphi$ and τ_{pp} [10, 42, 96–98].

To implement coherent-pump-probe spectroscopy, pulse pair generation by a simple beam splitter based delay line is not sufficient to independently control both τ_{pp} and $\Delta\varphi$. A more sophisticated approach using a pulse shaper or birefringent wedges becomes mandatory [10, 96, 99–103]. As a test to influence single molecule electronic coherences, Hildner and coworkers studied single TDI (terrylene diimide) molecules in a PMMA film with phase-locked double-pulse sequences at room temperature [10, 96]. In comparison to incoherent pump-probe experiments, which provide a phase averaged contribution of the electronic coherence (near zero-delay), the coherent scheme was found to reveal a fascinating interplay between electronic dephasing and population dynamics, even at room temperature. As discussed in Chaps. 7 and 8, such coherent superposition between two quantum states plays an important role in ultrafast biophotonics [30, 31].

Having described the basic phase control mechanism of an individual two-level system, the method can be expanded to excitation by broadband ultrashort pulses that cover several vibrational levels in a particular electronic transition. Such tracking of coherent molecular wavepacket evolution can yield insight into decoherence times as short as 30–40 fs [10, 104–107]. This scheme also forms the basis of coherent control discussed in Chap. 11. It enables targeted steering of a quantum system through phase shaped excitation pulses into one state from a multitude of accessible ones, while maintaining the potential of single molecule detection [108–110]. The coherent control of dynamic processes in molecular ensembles has been implemented in a variety of systems [111–113]. Fortunately, optimal control theory and experiments on ensembles have demonstrated that complex pulse shapes can often be simplified to physically more intuitive shapes based on trains of pulses with controlled width, delay and phase relations [10, 104–107]. Therefore, a pair of broadband ultrashort pulses with varying τ_{pp} and relative carrier envelope phase difference $\Delta\varphi$ (Chap. 2) can be used to observe vibrational wavepacket interference and phase control for an individual fluorophore at room temperature [10, 104–107].

4.6 Two-Colour Phase Control of Energy Transfer in Light Harvesting Complexes

Having set up the basic toolbox for coherent femtosecond single molecule spectroscopy, we now discuss one of its important applications in ultrafast biophotonics: ultrafast energy transfer in photosynthetic light harvesting complexes [28–31, 114, 115]. Detailed discussion of the structure of the photosynthetic complex and the energy transfer mechanisms is left for Chap. 7. The recent observation of room temperature quantum coherent dynamics in these photosynthetic systems has generated strong efforts, both in experiment and theory, to understand the role of quantum mechanics in biological functionalities [28–31, 114, 115]. So far, experimental approaches employed to investigate photosynthetic complexes have mainly concentrated on 2D electronic spectroscopy of ensembles of pigment-protein complexes [28, 29], yielding a spatial and temporal average over an inhomogeneous distribution. Consequently, the study of coherent ultrafast detection of individual complexes is vital to unravel the role of coherence in the light-harvesting efficiency of natural photosynthetic complexes.

Consider, by way of illustration, energy transfer in the light-harvesting 2 (LH2) complex of the purple photosynthetic bacteria, *Rhodopseudomonas acidophila* [10, 116]. The energy levels of the two LH2 fluophores, B800 and B850, along with the experimental scheme are depicted in Fig. 4.5. Individual LH2 complexes are excited by a two-colour pulse pair with delay τ_{pp} and phase difference $\Delta\varphi$ generated using a pulse shaper [10, 104–107]. The ensuing dynamics are probed by detecting the B850 spontaneous emission from a single complex using a confocal microscope [10, 116]. The first (blue) pulse coherently excites a band of vibrational levels in B800. Since the energy transfer between B800 and B850 involves quantum coherence, the excited state population of B850 maintains a certain phase relationship with the first pump pulse. The second (red) pump pulse, arriving after delay τ_{pp}, is in resonance with B850 and is also phase-locked ($\Delta\varphi$) with the first pulse. Therefore, the

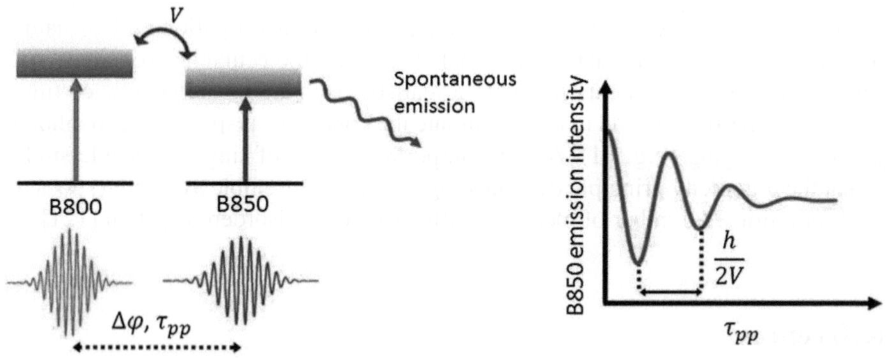

Fig. 4.5 Schematic of a two-colour phase control on single light-harvesting complex (LH2) containing fluophores B800 and B850 (*left*). A two-colour phase-locked pulse pair with delay τ_{pp} and phase difference $\Delta\varphi$ is used to coherently excite the two fluorphores. The first (*blue*) pulse induces excitation in the B800 band. In presence of coherent energy transfer (V) to the B850 band, the second (*red*) time-delayed pulse modulates the population transfer to B850 excited states via quantum interference. The resulting probability is probed by detecting the B850 spontaneous emission from a single complex. Time traces of the spontaneous emission show oscillatory intensity variations as a function of τ_{pp} reflecting quantum interference in population transfer

light-induced wavepacket in B850 can interfere with that generated via the energy transfer from B800 if τ_{pp} is less than the dephasing time of the complex [42, 97, 98]. Thus, the excited state population of B850 can be controlled coherently by varying τ_{pp} and $\Delta\varphi$. Since the intensity of spontaneous emission from B850 is proportional to the excited state population, its time evolution is a measure of the coherent population dynamics of the system. A periodic intensity variation as a function of τ_{pp} has been experimentally observed, which clearly reflects the quantum superposition of the excited states of the two fluophores [10, 116]. Intriguingly, coherence in LH2 is seen to persist for >300 fs, which is ~10 times longer than the coherence time of an isolated fluorophore in a solid state polymer environment! Thus, both the protein scaffold and the relatively strong electronic coupling between the fluophores play a key role in preserving coherence in the natural environment [117–120]. The interplay between persistent coherence and environment-assisted dissipation rapidly directs the excitation energy towards the lowest-energy target levels to prevent relaxation via competing loss channels [30, 31, 65]. The observed coupling constants and decoherence times indicate optimization such that the quantum coherences of the excited-state energy landscape survive long enough to avoid trapping [98], in local energy sinks (Chap. 7). Based on these observations, it is suggested that quantum coherence, both intrinsic and that induced by light, is essential in achieving the highest energy transfer efficiency in photosynthetic light-harvesting complexes [10, 116]!

Observation of electronic coherence, vibrational wave-packet interference, and coherent electronic energy transfer highlights the versatility of single-molecule ultrafast spectroscopy [9, 10, 16, 17]. From an ultrafast biophotonic viewpoint, the key aspect is that individual biomolecules are amenable to investigation in their natural surroundings. It is important to realize that, due to intermolecular heterogeneity, only single-molecule experiments facilitate a realistic estimate of the coherence times; ensemble experiments tend to underestimate the coherence response due to inhomogeneous averaging (Fig. 9.1). Another important aspect of single-molecule studies is that the ergodicity principle (time-averaging equals ensemble averaging) does not hold for a limited number of measurements made on a disordered system [121].

References

1. W. Siri, Rev. Sci. Instrum. **18**, 540 (1947)
2. W.E. Moerner, L. Kador, Phys. Rev. Lett. **62**, 2535 (1989)
3. M. Orrit, J. Bernard, Phys. Rev. Lett. **65**, 2716 (1990)
4. W.E. Moerner, M. Orrit, Science **283**, 1670 (1999)
5. J.J. Macklin, J.K. Trautman, T.D. Harris, L.E. Brus, Science **272**, 255 (1996)
6. J.A. Veerman, M.F. Garcia-Parajo, L. Kuipers, N.F. van Hulst, Phys. Rev. Lett. **83**, 2155 (1999)
7. K.D. Weston, P.J. Carson, H. Metiu, S.K. Buratto, J. Chem. Phys. **109**, 7474 (1998)
8. R.A.L. Vallée, N. Tomczak, L. Kuipers, G.J. Vancso, N.F. van Hulst, Phys. Rev. Lett. **91**, 38301 (2003)
9. M. Orrit, T. Ha, V. Sandoghdar, Chem. Soc. Rev. **43**, 973 (2014)
10. D. Brinks, R. Hildner, E.M.H.P. van Dijk, F.D. Stefani, J.B. Nieder et al., Chem. Soc. Rev. **43**, 2476 (2014)
11. A. Yildiz, J.N. Forkey, S.A. McKinney, T. Ha, Y.E. Goldman, P.R. Selvin, Science **300**, 2061 (2003)
12. S. Schlücker, Angew. Chem. Int. Ed. **53**, 4756 (2014)
13. W.E. Moerner, J. Phys. Chem. B **106**, 910 (2002)
14. S. Weiss, Science **283**, 1676 (1999)
15. B.N.G. Giepmans, S.R. Adams, M.H. Ellisman, R.Y. Tsien, Science **312**, 217 (2006)
16. B. Schuler, H. Hofmann, Curr. Opin. Chem. Biol. **23**, 36 (2013)
17. M.F. Juette, D.S. Terry, M.R. Wasserman, Z. Zhou, R.B. Altman et al., Curr. Opin. Chem. Biol. **20**, 103 (2014)
18. K.I. Willig, S.O. Rizzoli, V. Westphal, R. Jahn, S.W. Hell, Nature **440**, 935 (2006)
19. S.W. Hell, Science **316**, 1153 (2007)
20. G.R. Fleming, *Chemical Applications of Ultrafast Spectroscopy* (Oxford University Press, New York, 1986)
21. O. Svelto, S. De Silvestri, G. Denardo, *Ultrafast Processes in Spectroscopy* (Plenum Press, New York, 1996)
22. M. Braun, P. Gilch, W. Zinth, *Ultrashort Laser Pulses in Biology and Medicine* (Springer, Berlin, 2008)
23. C. Manzoni, D. Polli, G. Cerullo, Rev. Sci. Instrum. **77**, 023103 (2006)
24. D. Polli, L. Luer, G. Cerullo, Rev. Sci. Instrum. **78**, 103108 (2007)
25. P. Vasa, W. Wang, R. Pomraenke, M. Lammers, M. Maiuri et al., Nat. Photon. **7**, 128 (2013)
26. S. Mukamel, Anu. Rev. Phys. Chem. **51**, 691 (2000)
27. M. Zanni, R.M. Hochstrasser, Curr. Opin. Struct. Biol. **11**, 516 (2001)
28. G.S. Engel, T.R. Calhoun, E.L. Read, T.-K. Ahn, T. Mančal et al., Nature **446**, 782 (2007)
29. A. Ishizaki, T.R. Calhoun, G.S. Schlau-Cohen, G.R. Fleming, Phys. Chem. Chem. Phys. **12**, 7319 (2010)
30. N. Lambert, Y.-N. Chen, Y.-C. Cheng, C.-M. Li, G.-Y. Chen, F. Nori, Nat. Phys. **9**, 10 (2013)

31. M. Mohseni, Y. Omar, G.S. Engel, M.B. Plenio (eds.), *Quantum Effects in Biology* (Cambridge University Press, Cambridge, 2014)
32. Y. Zhang, Y.R. Zhen, O. Neumann, J.K. Day, P. Nordlander, N.J. Halas, Nat. Commun. **5**, 4424 (2014)
33. M. Fleischmann, P.J. Hendra, A.J. McQuillan, Chem. Phys. Lett. **26**, 163 (1974)
34. R. Hillenbrand, T. Taubner, F. Keilmann, Nature **418**, 159 (2002)
35. A. Hartschuh, E.J. Sánchez, X.S. Xie, L. Novotny, Phys. Rev. Lett. **90**, 095503 (2003)
36. J. Steidtner, B. Pettinger, Phys. Rev. Lett. **100**, 236101 (2008)
37. R. Zhang, Y. Zhang, Z.C. Dong, S. Jiang, C. Zhang et al., Nature **498**, 82 (2013)
38. M. Agio, A. Alù, *Optical Antennas* (Cambridge University Press, Cambridge, 2013)
39. T.W. Ebbesen, H.J. Lezec, H.F. Ghaemi, T. Thio, P.A. Wolff, Nature **391**, 667 (1998)
40. W.L. Barnes, A. Dereux, T.W. Ebbesen, Nature **424**, 824 (2003)
41. C. Ropers, D.J. Park, G. Stibenz, G. Steinmeyer, J. Kim et al., Phys. Rev. Lett. **94**, 113901 (2005)
42. R.W. Boyd, *Nonlinear Optics*, 3rd edn. (Academic Press, London, 2008)
43. T. Ichimura, N. Hayazawa, M. Hashimoto, Y. Inouye, S. Kawata, Phys. Rev. Lett. **92**, 220801 (2004)
44. J.A. Fan, K. Bao, C. Wu, J. Bao, R. Bardhan et al., Nano Lett. **10**, 4680 (2010)
45. J. Ye, F. Wen, H. Sobhani, J.B. Lassiter, P. van Dorpe et al., Nano Lett. **12**, 1660 (2012)
46. Y. Zhang, F. Wen, Y.R. Zhen, P. Nordlander, N.J. Halas, Proc. Nat. Acad. Sci. USA **110**, 9215 (2013)
47. K. Thyagarajan, J. Butet, O.J.F. Martin, Nano Lett. **13**, 1847 (2010)
48. C. Joo, H. Balci, Y. Ishitsuka, C. Buranachai, T. Ha, Ann. Rev. Biochem. **77**, 51 (2008)
49. M. Rief, Curr. Opin. Chem. Biol. **23**, 48 (2013)
50. P.R. Selvin, T. Ha, *Single-Molecule Techniques: A Laboratory Manual* (Cold Spring Harbor Laboratory Press, New York, 2008)
51. T. Ha, P. Tinnefeld, Ann. Rev. Phys. Chem. **63**, 595 (2012)
52. Q. Zheng, M.F. Juette, S. Jockusch, M.R. Wasserman, Z. Zhou et al., Chem. Soc. Rev. **43**, 1044 (2014)
53. S. Hohng, C. Joo, T. Ha, Biophys. J. **87**, 1328 (2004)
54. H. Kim, T. Ha, Rep. Progr. Phys. **76**, 016601 (2013)
55. S. Hohng, S. Lee, J. Lee, M.H. Jo, Chem. Soc. Rev. **43**, 1007 (2014)
56. H. Gelman, M. Platkov, M. Gruebele, Chemistry **18**, 6420 (2012)
57. Y. Phillip, V. Kiss, G. Schreiber, Proc. Nat. Acad. Sci. USA **109**, 1461 (2012)
58. M.A. Medina, P. Schwille, BioEssays **24**, 758 (2002)
59. D. Nettels, I.V. Gopich, A. Hoffmann, B. Schuler, Proc. Nat. Acad. Sci. USA **104**, 2655 (2007)
60. D. Nettels, A. Hoffmann, B. Schuler, J. Phys. Chem. B **112**, 6137 (2008)
61. I.V. Gopich, D. Nettels, B. Schuler, A. Szabo, J. Chem. Phys. **131**, 095102 (2009)
62. A. Soranno, B. Buchli, D. Nettels, S. Müller-Späth, R.R. Cheng et al., Proc. Nat. Acad. Sci. USA **109**, 17800 (2012)
63. J.R. Lakowicz, *Principles of Fluorescence Spectroscopy*, 3rd edn. (Springer-Verlag, Berlin, 2006)
64. W. Becker, *Advanced Time-Correlated Single Photon Counting Techniques* (Springer-Verlag, Berlin, 2005)
65. V. Sundström, Annu. Rev. Phys. Chem. **59**, 53 (2008)
66. M. Dantus, R.M. Bowman, A.H. Zewail, Nature **343**, 737 (1990)
67. X. Hu, K. Schulten, Phys. Today **28**, (1997)
68. P.B. Corkum, F. Krausz, Nat. Phys. **3**, 381 (2007)
69. M. Merano, S. Sonderegger, A. Crottini, S. Collin, P. Renucci et al., Nature **438**, 479 (2005)
70. V.A. Lobastov, J. Weissenrieder, J. Tang, A.H. Zewail, Nano Lett. **7**, 2552 (2007)
71. B.J. Siwick, J.R. Dwyer, R.E. Jordan, R.J.D. Miller, Science **302**, 1382 (2003)
72. H. Ihee, V.A. Lobastov, U.M. Gomez, B.M. Goodson, R. Srinivasan et al., Science **291**, 5503 (2001)
73. C. Rose-Petruck, R. Jimenez, T. Guo, A. Cavalleri, C.W. Siders et al., Nature **398**, 310 (1999)

74. C.W. Siders, A. Cavalleri, K. Sokolowski-Tinten, C. Toth, T. Guo et al., Science **286**, 1340 (1999)
75. M. Bargheer, N. Zhavoronkov, Y. Gritsai, J.C. Woo, D.S. Kim et al., Science **306**, 1771 (2004)
76. P. Vasa, C. Ropers, R. Pomraenke, C. Lienau, Laser Photon. Rev. **3**, 483 (2009)
77. B. Hecht, L. Novotny, *Principles of Nano-optics* (Cambridge University Press, Cambridge, 2005)
78. M. Ohtsu, K. Kobayashi, T. Kawazoe, T. Yatsui, M. Naruse, *Principles of Nanophotonics* (CRC Press, Boca Raton, 2007)
79. E. Betzig, J.K. Trautman, T.D. Harris, J.S. Weiner, R.L. Kostelak, Science **251**, 1468 (1991)
80. E. Betzig, J.K. Trautman, Science **257**, 189 (1992)
81. F. Zenhausern, Y. Matin, H.K. Wickramasinghe, Science **269**, 1083 (1995)
82. B. Knoll, F. Keilmann, Nature **399**, 134 (1999)
83. R. Müller, C. Lienau, App. Phys. Lett. **76**, 3367 (2000)
84. K. Matsuda, T. Saiki, S. Nomura, M. Mihara, Y. Aoyagi, App. Phys. Lett. **81**, 2291 (2002)
85. L. Novotny, R.X. Bian, X.S. Xie, Phys. Rev. Lett. **79**, 645 (1997)
86. E.J. Sánchez, L. Novotny, X.S. Xie, Phys. Rev. Lett. **82**, 4014 (1999)
87. A. Bouhelier, M. Beversluis, A. Hartschuh, L. Novotny, Phys. Rev. Lett. **90**, 013903 (2003)
88. C.C. Neacsu, G. Steudle, M.B. Raschke, App. Phys. B **80**, 295 (2005)
89. S. Amarie, F. Keilmann, Phys. Rev. B **83**, 045404 (2011)
90. M. Brehm, A. Schliesser, F. Keilmann, Opt. Express **14**, 11222 (2006)
91. C. Ropers, D.R. Solli, C.P. Schulz, C. Lienau, T. Elsaesser, Phys. Rev. Lett. **98**, 043907 (2007)
92. C. Ropers, C.C. Neacsu, T. Elsaesser, M. Albrecht, M.B. Raschke, C. Lienau, Nano Lett. **7**, 2784 (2007)
93. B.A. Nechay, U. Siegner, M. Achermann, H. Bielefeldt, U. Keller, Rev. Sci. Instrum. **70**, 2758 (1999)
94. R. Pomraenke, C. Ropers, J. Renard, C. Lienau, L. Lüer et al., Nano Lett. **7**, 998 (2007)
95. D. Davydova, A. de la Cadena, D. Akimov, B. Dietzek, Laser Photon. Rev. **10**, 62 (2016)
96. R. Hildner, D. Brinks, N.F. van Hulst, Nat. Phys. **7**, 172 (2011)
97. M.O. Scully, M.S. Zubairy, *Quantum Optics* (Cambridge University Press, Cambridge, 2001)
98. G. Lanzani, *The Photophysics Behind Photovoltaics and Photonics* (Wiley-VCH, Weinheim, 2012)
99. M.L. Cowan, J.P. Ogilvie, R.J.D. Miller, Chem. Phys. Lett. **386**, 184 (2004)
100. T. Brixner, I.V. Stiopkin, G.R. Fleming, Opt. Lett. **29**, 884 (2004)
101. T. Zhang, C.N. Borca, X.Q. Li, S.T. Cundiff, Opt. Express **13**, 7432 (2005)
102. U. Selig, F. Langhojer, F. Dimler, T. Löhrig, C. Schwarz et al., Opt. Lett. **33**, 2851 (2008)
103. D. Brida, C. Manzoni, G. Cerullo, Opt. Lett. **37**, 3027 (2012)
104. D. Brinks, F.D. Stefani, F. Kulzer, R. Hildner, T.H. Taminiau et al., Nature **465**, 905 (2010)
105. D. Brinks, R. Hildner, F.D. Stefani, N.F. van Hulst, Faraday Disc. **153**, 51 (2011)
106. D. Brinks, R. Hildner, F.D. Stefani, N.F. van Hulst, Opt. Express **19**, 26486 (2011)
107. N. Accanto, J.B. Nieder, L. Piatkowski, M. Castro-Lopez, F. Pastorelli et al., Light: Sci. Appl. **3**, e143 (2014)
108. D.J. David, S.A. Rice, J. Chem. Phys. **83**, 5013 (1985)
109. P. Brumer, M. Shapiro, Chem. Phys. Lett. **126**, 541 (1986)
110. R.J. Gordon, S.A. Rice, Ann. Rev. Phys. Chem. **48**, 601 (1997)
111. M. Dantus, V.V. Lozovoy, Chem. Rev. **104**, 1813 (2004)
112. D. Meshulach, Y. Silberberg, Nature **396**, 239 (1998)
113. Y. Silberberg, Annu. Rev. Phys. Chem. **60**, 277 (2009)
114. M. Sarovar, A. Ishizaki, G.R. Fleming, K.B. Whaley, Nat. Phys. **6**, 462 (2010)
115. D. Scholes, Nat. Phys. **6**, 402 (2010)
116. R. Hildner, D. Brinks, J.B. Nieder, R.J. Cogdell, N.F. van Hulst, Science **340**, 1448 (2013)
117. E. Collini, C.Y. Wong, K.E. Wilk, P.M.G. Curmi, P. Brumer, G.D. Scholes, Nature **463**, 644 (2010)
118. G. Panitchayangkoon, D. Hayes, K.A. Fransted, J.R. Caram, E. Harel et al., Proc. Nat. Acad. Sci. USA **107**, 12766 (2010)
119. A. Ishizaki, G.R. Fleming, J. Phys. Chem. B **115**, 6227 (2011)
120. C.Y. Wong, R.M. Alvey, D.B. Turner, K.E. Wilk, D.A. Bryant et al., Nat. Chem. **4**, 396 (2012)
121. P. Walters, *An Introduction to Ergodic Theory* (Springer, New York, 1982)

Chapter 5
Ultrafast Lasers in Surgery and Cell Manipulation

Abstract In this chapter we present an overview of two applications that are already very well established: laser surgery and cell manipulation with laser light. We explore here some of the essential physics that drives the carrying out of surgery and single cell manipulation using femtosecond-duration laser pulses.

5.1 Introduction

Light has been used by biologists to illuminate a variety of entities for microscopic investigations over the decades. The intensity of light in such applications is so low that it precludes any significant influence on the sample that is being observed. With the advent of lasers as sources of light, focused illumination can often attain intensities that are strong enough to begin influencing the physio-chemical properties of matter that is under observation. Indeed, intense enough light can now be utilized to optically *manipulate* matter, including living entities. In this context, biologists encounter an entire gamut of possibilities at different length scales:

(i) At the level of single biomolecules it is possible to utilize strongly focused continuous wave light to apply forces and torques [1];
(ii) At the sub-cellular level, it is now possible to toggle individual biomolecules from an active state to a passive state—and vice versa—by photoswitching of fluorescent labels [2];
(iii) At the cellular level, light has been used to activate membrane channels and membrane pumps in order to affect alterations of electrical potential across cell membranes [3], enabling neural circuitry to begun to be experimentally probed; and
(iv) At the tissue level, ablation induced by pulsed lasers has allowed the cutting of diffraction limited volumes of tissue and to facilitate transection of cells [4].

Out of these, the ultrafast aspects of light-matter interactions manifest themselves most prominently in the last category of encounters, and it is this that we shall focus on in this chapter.

P. Vasa and D. Mathur, *Ultrafast Biophotonics*, Biological and Medical Physics, Biomedical Engineering, DOI 10.1007/978-3-319-39614-9_5

The use of pulsed lasers, specially those that produce femtosecond-duration pulses, permits the attainment of high instantaneous peak power that facilitates the onset of nonlinear (multiphoton) absorption while, concomitantly, avoiding heat damage to the irradiated sample via linear absorption. The latter results from the *average* power from such lasers being very low and the wavelength of the laser light generally being non-absorbing by biomolecules. Hence, in the course of the last decade or so femtosecond lasers have begun to be of considerable utility in an entire gamut of non-surgical applications, like multiphoton laser scanning microscopy with utility in the neurosciences [5, 6], high harmonic imaging of cells, tissues and organisms [7–9], coherent anti-Stokes Raman spectroscopy [10], and nonlinear imaging of biological entities in general [11].

Pulsed laser irradiation can lead to formation of plasma. As is discussed in other contexts throughout the rest of this book, femtosecond laser-induced plasma formation occurs in two-step fashion:

(1) Valence shell electrons undergo a bound→free transition due to multiphoton and tunnel effects (Chap. 2);
(2) The ionized electrons "feel" the effect of the optical field via inverse bremsstrahlung (IB). These electrons are accelerated by the ponderomotive force (Chap. 2); the optical field reverses direction after half a cycle and this leads the electrons to now be accelerated in the opposite direction: towards their ionic progenitors. Rescattering of these electrons with the progenitors can give rise to even more electrons. As is shown in Chaps. 2, 6 and 11, this electron rescattering process takes place on sub-femtosecond timescales and it also occurs in an aqueous environment—notwithstanding relatively slow-acting solvation effects—and is, consequently, of importance in a biological context. These two steps, leading to further ionization and continued generation of ionized electrons, can drive the growth of micrometre-sized plasma bubbles. With sufficient electron density, the plasma conductivity finally limits the penetration of the incident laser radiation to a very small skin depth, of the order of a few tens of nanometres. Such axial localization of the plasma is far less than typical Rayleigh (confocal) lengths of the incident radiation.

After the pulse duration, the ionized electrons can recombine with the positively charged ions within the focal volume, typically on picosecond timescales. The resulting transfer of the electrons' kinetic energy to proximate material occurs on timescales that are shorter than the acoustic relaxation time of the material (typically a few hundred picoseconds). Such impulsive release of energy can rupture the ambient material via formation of cavitation bubbles (there is further discussion on bubbles in Chap. 11). Such bubbles lie at the heart of laser-induced damage—or ablation—in the context of biophotonics.

As is indicated in Fig. 5.1, a 100 fs laser pulse of 10 nJ energy, as would be readily obtained from a Ti:sapphire oscillator, can yield a fluence of 1 J cm^{-2} upon being focused to a 1 μm^2 spot. The resulting intensity exceeds 10 TW cm^{-2} during most of the 100 fs, a value which is more than the threshold for ionization and, thence, plasma formation. Note that a two-order-of-magnitude larger value of total pulse energy

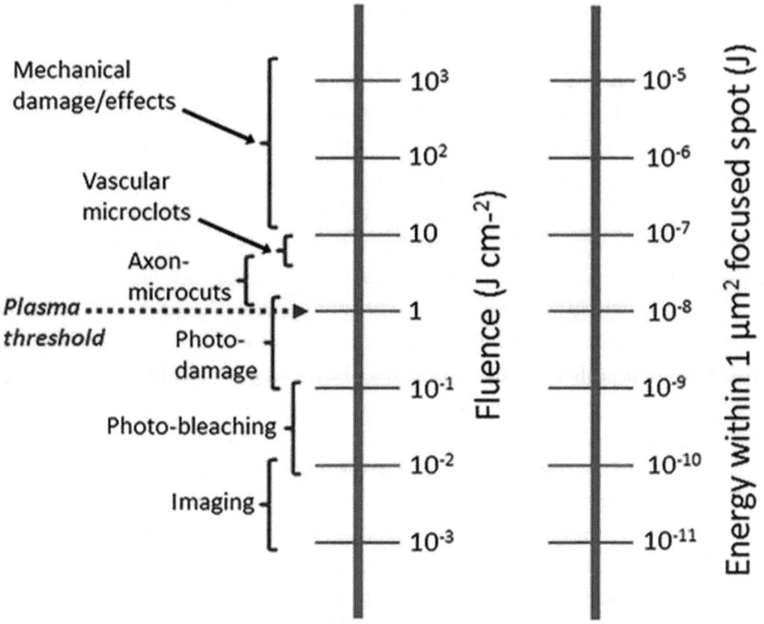

Fig. 5.1 An overview of some scales of laser fluence of relevance in laser surgery. A 10 nJ laser beam (of 100 fs pulses) focused to a spot size of 1 μm^2 results in a fluence of 1 J cm^{-2}; this value constitutes the approximate threshold for plasma formation in a physiological environment. Lower values of fluence are of utility in imaging and for localized photodamage and photobleaching

would be required to overcome the plasma threshold in the case of a picosecond pulse. In such cases, the very much higher total pulse energy would most certainly cause collateral damage to proximate tissue due to heating, bubble formation and expansion while the pulse is still being applied, and the consequent release of shockwaves. It is for this reason that ultrafast (femtosecond-duration) laser pulses with fluences close to the 1 J cm^{-2} threshold are preferred for laser surgery applications.

Considerable progress has been achieved in the last decade on application of femtosecond-duration pulses in biophotonic applications like sub-micrometre surgery of neuronal and vascular entities [12] and in various types of eye surgery [13–16]. It is of interest to draw attention to a study that was conducted in Mexico on the incidence of intra-operative flap complications during eye surgery [17]. Typically, 1053 nm light pulses from a femtosecond laser are focused to a ~3 μm spot, with a typical accuracy of 1 μm, to cut a spiral pattern in the corneal stroma, thereby forming precise lamellar flaps for surgery. The Mexican study investigated 26,600 patients who had undergone such eye surgery. Out of these, only 65 patients reported any complications (constituting 0.24 % of the total sample) [17]. An impressive 99.76 % (26,535 patients) had uneventful flap creation using a femtosecond laser beam.

5.2 Nanosurgery and Manipulation of Cells and Tissues

Biological media are usually transparent to 800 nm light that emanates from various laser sources (like Ti:sapphire lasers, fibre lasers, semiconductor lasers). This statement holds true at low values of irradiance. However, when the laser light is focused through a microscope objective with high numerical aperture (NA), it became evident as long ago as the 1980s that highly localized laser-induced effects can begin to manifest themselves [18, 19]. This realization gave advent to the use of nanosecond laser pulses, in conjunction with microscope objectives of moderate NA, to carry out intra-ocular surgery [20]. Subsequent use of femtosecond lasers widened the scope in very considerable fashion, with the widespread advent of LASIK (Laser-Assisted in situ Keratomileusis) allowing diverse types of eye surgeries to be performed, ranging from the creation of corneal flaps [21–24] to various corneal intra-stromal refractive surgeries [23, 24].

The physics that underpinned such developments came to be unraveled once it was realized that even when microscope objectives of moderate NA were used, a host of nonlinear propagation processes governed how the incident laser energy was spatially distributed on the target. Prominent among such nonlinear processes was filamentation, caused by the interplay of Kerr self-focusing, group velocity dispersion, and plasma defocusing [25–29]. We shall discuss these processes later in this chapter and shall have occasion to invoke them in other parts of the book.

Linear optical processes like diffraction that may occur at apertures might also play a contributory role [30] and it is, therefore, important to consider the following practical points. Nonlinear propagation effects manifest themselves more strongly when shorter laser pulses are used. However, the use of ultrashort pulses with low NA optics makes it difficult to deposit optical energy in highly localized fashion, which is what applications like surgery require. Hence, focusing at large values of NA becomes mandatory from the practical expediency of minimizing the effects of filamentation as also making the diffraction-limited focal volume as small as possible. We note that while Kerr self-focusing (Chap. 2) requires a certain threshold power to be surpassed, it is the irradiance threshold that needs to be surpassed for optically-induced breakdown (plasma formation) to occur. Large values of NA enable the focused volume to be made smaller and, correspondingly, less incident laser power is required to surpass the irradiance threshold. Beyond a certain high value of NA, the threshold for optically-induced breakdown becomes less than the critical power for Kerr self-focusing. Under such circumstances, it becomes possible to attain energy deposition on highly localized (sub-μm) scales. In the case of femtosecond laser pulses (1 mJ m^{-2}), optically-induced breakdown in water and glass occurs for NA values equal to, or in excess of, 0.9 [31].

In addition to surgical applications, interest in light-matter interactions at the cellular and sub-cellular level predates, by many decades, the advent of ultrafast pulsed lasers. As long ago as 1912 the German physicist turned prominent anti-fascist, S. Tschachotin, made use of tightly focused ultraviolet light to carry out optically-induced inactivation of cells and cell organelles [32, 33]. The light used in these pioneering experiments was of 280 nm wavelength, obtained from a magnesium spark; it was focused using a microscope objective to a 5 μm diameter spot on a cell placed on a microscope slide. Of necessity, very long exposure times were required before any optically-induced modification of cellular properties or structure would occur. The use of pulsed laser sources opened new prospects. The use of argon ion lasers in the 1980s enabled exposure times to be shortened but, it was soon realized, the energy deposition relied essentially on linear absorption of a single photon. This, in turn, placed certain practical restrictions. For instance, the biological sample of interest had to be stained with chemical dyes or, instead, laser wavelengths had to be employed that would be absorbed by the biological sample in its pristine state. These restrictions, which persist in a wide variety of circumstances and for a large subset of biological matter, can be an important consideration if one is interested in optically inducing alterations within a cell or an organelle with a view to studying how specific proteins interact with cellular matter in the course of biological processes like embryonic development and cell proliferation or in the mapping of stress-induced reaction pathways.

The use of short-pulse (femtosecond) lasers has enabled progress to be achieved in such functional studies, mainly due to the adoption of one of two different experimental strategies. In the so-called local approach [34] a tightly focused femtosecond laser beam is utilized to expose within the focal volume one (or very few) target cells which can be physically dissected or chemically altered. The strategy is to make use of high enough local laser intensities to be generated such that nonlinear absorption occurs within the focal volume: cell surgery can thus be performed at any location within the cell volume independently of its linear absorption properties.

In an alternative, systemic approach, a specific protein is made the target of the focused laser beam; the protein is accessed by means of appropriate antibodies that are attached to nanoparticles or chromophores [35–38]. Protein inactivation occurs upon linear absorption of optical energy by the nanoparticles or chromophores and can result in thermo-mechanical or photochemical destruction of the protein host. In this technique the spatial location of the nanoparticles or chromophores within the cell is immaterial and, in that sense, this is a non-local approach. A cogent description of how these two approaches have been used in cellular and sub-cellular surgery with nanosecond, picosecond and femtosecond lasers has been presented in a masterful review by Vogel and coworkers [34].

There are a plethora of other manifestations of cellular surgery that have been developed in the course of the decade since femtosecond laser pulses first began to be employed.

It is established that analytical techniques like polymerase chain reaction (PCR), including time-dependent variants, are now the mainstay of investigations of cell constituents and their specific biochemical functions. PCR facilitates the analysis of minute quantities of biomaterials, including cells. However, these need to be first separated from tissue samples that are heterogeneous and ultrafast laser techniques are now coming into their own in this domain. The radiation pressure of an intense laser beam has been utilized for laser pressure catapulting (LPC) of dissected samples into appropriate containers for PCR studies [39]. A related use of ultrashort laser beams has been to carry out cell lysis and subsequent catapulting of the contents of the lysed cell into a micropipette for time-resolved capillary electrophoresis [40], for fusing together of cells as well as for in-vitro fertilization wherein the laser beam is used to ablate the outer egg membrane [41]. In recent years less vigorous processes have been induced by ultrafast laser beams, such as the gentle transfection of materials like genes into specific cells, wherein the tightly focused laser beam acts as a sharp "needle" (see [42–45] and references therein).

There have been very diverse applications of femtosecond and longer laser pulses in surgical and cell manipulation studies. Amongst the earliest of the latter type of studies succeeded in probing how morphogenetic motion occurs in the course of embryonal development [46], how nerves regenerate after nerve cells are cut in worms [47], and how waves of calcium ions can be stimulated in living cells [48]. Several early studies focused on studying cell division, especially on the separation of chromosomes [49–52], and on quantifying the biophysical properties of the cytoskeleton and mitochondria [53]. A key question faced by the early researchers concerned the relative advantages of femtosecond and picosecond laser pulses.

With picosecond laser pulses researchers were able to achieve intracellular dissection using energies of the order of 100 nJ [50, 52]. Later, it was found that similar dissection could be achieved with energies that were two orders of magnitude less (about 1 nJ) when femtosecond pulses were employed [47]. These ultrashort pulses create plasma at lower energy thresholds and can induce very fine effects extending over spatial volumes that may be smaller than the optical diffraction limit. This advantage was quickly recognized and put to good use in studies covering a very large range of sizes, from tissues [44, 54, 55] and small organisms [46, 47] to various organelles [53, 56, 57], including chromosomes [58, 59], and experiments on localized damage induced in DNA [60]. It was appreciated that a critical factor that underlines differences between femtosecond pulses on the one hand and longer-duration pulses on the other is the nature of the energy absorption process. For nanosecond pulses, the threshold for optical breakdown is dependent on how efficiently optical energy is absorbed by the bio-sample in the focal volume. This is the domain of linear absorption and the parameter of importance is the linear absorption coefficient of the bio-sample that is being irradiated. With the use of near-IR femtosecond pulses, the material's absorption coefficient ceases to be of primary importance because nonlinear absorption dominates the transfer of energy from the optical field into the bio-sample. Concomitantly, the nonlinearity ensures that the effective volume within which efficient energy absorption occurs is much smaller than in the case of longer-duration pulses. This affords a distinct advantage to femtosecond pulses: smaller targets, such as cellular assemblies, can be efficiently irradiated.

Biomolecules will, of course, readily absorb ultraviolet radiation and it is, therefore, not surprising that nitrogen lasers also found utility in early studies involving micro-dissection of tissues. UV pulses offer the advantage of much lower ablation thresholds than those obtained with visible or near-IR pulses of the same fluence. It has been shown that UV laser-induced micro-dissection occurs as a consequence of linear absorption followed by plasma formation followed by the onset of mechanical effects like the generation of shock waves and cavitation bubble formation [18].

A disadvantage is that the mechanical effects are seldom localized and, consequently, the use of nanosecond long pulses of UV light of even modest energy (μJ range) can have deleterious effects on cell viability. On the other hand, femtosecond pulses of near-IR light yield a low energy threshold for plasma formation [61]. Use of such pulses has enabled fine dissection to be carried out over spatial extents that are smaller than the optical diffraction limit with only nanoJoules of incident laser energy in a variety of chromosomes, cell organelles, tissues as well as small organisms [19, 23, 46, 47, 54–56]. As noted earlier, the sub-diffraction-limited spatial effects are a consequence of nonlinear absorption limiting the volume into which the incident laser energy gets deposited. A concomitant advantage is that nonlinear modification of the biological sample can be readily imaged using nonlinear techniques such as second harmonic generation and two-photon fluorescence excitation [43, 46, 47, 54, 57] (Chap. 3).

Another major step that contributed immensely to progress in the field was the adoption of gene fusion products to aid microscopic visualization. Green fluorescent proteins (GFPs) are the most well known example, and they have allowed visualization and ablation of cellular structures at unprecedently high values of resolving power [46, 47, 62], certainly higher than resolving powers readily achieved with conventional light microscopy.

5.3 Optical Field Ionization: Plasma Formation, Energetics and Temporal Considerations

Plasma produced upon irradiation by femtosecond laser pulses is the main driver for nanosurgical applications and for manipulation of transparent biomaterials. Plasma is also, potentially, the agent responsible for sample damage in applications such as high harmonic imaging and in multiphoton microscopy (Chap. 3). It is, therefore, pertinent to present a brief overview of plasma formation processes that occur upon femtosecond laser irradiation of biomaterials. For purposes of illustration we shall focus our discussion on plasma formation in the most common and important of biological constituents, aqueous water.

A key step in plasma formation is, of course, the formation of free electrons in the laser-induced ionization process. In the condensed phase, "free" electrons are really only quasi-free in the sense that ionization implies a transition from the valence band to the conduction band. Water has been successfully modelled as an amorphous semiconductor, with a bandgap of 6.5 eV [63]. Although there are many models of solvation and thermalization dynamics of electrons produced upon multiphoton ionization of aqueous H_2O, there has been limited success in rationalizing experimental observations (see, for instance, [64–68], and references therein). Nevertheless, it seems established that upon irradiation by an intense beam of laser light, multiphoton excitation and ionization of H_2O occurs. With low-energy (<9 eV) photons, electrons that are ejected in the ionization process are expected to be $\sim$1 nm away from their parent holes. On the other hand, with higher energy (>11 eV) photons, the photoelectrons may be as far as 3 nm away [64]. In both cases the photoionization process probably involves a sequence of proton transfers and electron transfers to pre-existing traps although, it is most likely that higher energy photoexcitation leads to electrons being ejected directly into the conduction band of water. Under strong-field conditions, that is when incident laser intensities are in the 10 TW cm^{-2} range and above, it is the tunnel ionization process (Chap. 2) that is expected to dominate and, under such circumstances, direct transfer to the conduction band is the most likely scenario that governs electron dynamics.

Another facet to consider when dealing with ionization in the condensed phase is the effect of solvation. Water possesses a naturally high polarity; this ensures that a large percentage of electrons produced in the ionization process become solvated. Solvation is said to occur when the proximate H_2O molecules form a cage which "traps" the ionized electrons by acting as a screen of the free charge. It has been estimated that such solvated states live for period that are as long as ~500 ps [64–68].

The effects related to propagation of intense femtosecond laser pulses through the liquid water also need to be taken into consideration in understanding the overall dynamics. Filaments (or plasma channels) can form, and refractive index variations can occur which can, subsequently, lead to transient absorption. The shadowgraphy technique has been utilized to characterize the spatial and temporal variation of refractive index and transient absorption induced by a filament generated as an intense 120 fs long laser pulse propagates in water [69]: electron densities as high as $\sim 10^{18}$ cm^{-3} are shown to be readily achieved in liquid water.

Upon reaching the conduction band, the ionized electrons are subjected to acceleration by the ponderomotive potential when intense enough laser pulses are utilized. Electron kinetic energies in excess of 5 eV can be reached upon irradiation by 1350 nm light at 25 TW cm^{-2} intensity; corresponding values at the same incident intensity are as high as 9 eV in the case of longer wavelength, 2200 nm, light. Single photon energies corresponding to other wavelengths that are employed are as follows: 1064 nm (1.17 eV), 800 nm (1.56 eV), 532 nm (2.34 eV), and 355 nm (3.51 eV). These values demand that 6, 5, 3, and 2 photons, respectively, are needed to overcome the 6.5 eV bandgap in water. For laser pulses in excess of a few hundred femtosecond duration, it is probably a fair assumption that ionization of water occurs the instant 6.5 eV worth of energy is supplied. The situation becomes somewhat modified when even shorter laser pulses are utilized. Now, the effective value of the bandgap is that which takes account of the oscillation energy of the ionized electron as it continues to experience the optical field. If we denote the bandgap as Δ, then the ionization energy can be expressed in terms of the Keldysh formulation (Chap. 2) as

$$\Delta_{eff} = \Delta + \frac{e^2 E^2}{4m\omega^2}, \tag{5.1}$$

where E, ω denote the optical field's amplitude and angular frequency, respectively, e is the electronic charge, and m is the exciton reduced mass:

$$\frac{1}{m} = \frac{1}{m_c} + \frac{1}{m_v}, \tag{5.2}$$

with m_c being the effective mass of the ionized (quasi-free) electron in the conduction band and m_v being the mass of the hole in the valence band. It is clear why $\Delta_{eff} \sim \Delta$ for long pulses: it is only with ultrashort pulses that the magnitude of the optical field, E, becomes large enough to be an important factor.

As already noted, the quasi-free electron produced upon ionization continues to experience the optical field and one of the consequences is that it can absorb photons via a non-resonant IB process. In addition to the electron and the photon, the presence of a third particle, like the H_2O^+ ion, is necessary in order to conserve energy and momentum in the course of the IB process. We have also noted that a second consequence will be that the ionized electron undergoes a reversal of its initial post-ionization trajectory after half an optical cycle, once the direction of the laser field changes. A sequence of such IB events is that an electron may gain enough kinetic energy so as to be able to induce further ionization, this time via a simple electron-molecule collision that occurs due to laser-field-induced rescattering (Chap. 2). It is not difficult to imagine that a sequence of such events can lead to the possibility of avalanche ionization: this can occur if the incident laser intensity is high enough to overcome the loss of quasi-free electrons from the focal volume—via diffusion processes, or by recombination. Although it is known that electron-ion recombination cross sections are very low in the case of atomic ions—the recombination then occurs mostly by radiative processes and, even more rarely, by a dielectronic process—the cross sections can be very large when the interaction is between low energy electrons and *molecular* ions like H_2O^+ or OH^+. The latter process is known as dissociative recombination and cross sections for such recombination events can be as large as 10^{-14} cm^2 for very low electron energies [70] as would be encountered in aqueous environments of biological relevance. Electron depletion will certainly occur very efficiently when dissociative recombination occurs. Hence, avalanche ionization would require the rate of energy gain by quasi-free electrons via IB to proceed much more rapidly than the rate at which dissociative recombination occurs.

In order to gain some further insight into the overall energetics, let us also consider the following. Collisional ionization is integral to the avalanche ionization process. For a quasi-free electron to induce further ionization, it must possess kinetic energy in excess of the effective ionization energy, Δ_{eff}, referred to above. In the condensed phase, it is known that the critical energy, E_{crit}, for bands possessing parabolic energy dispersion can be deduced using [71]

$$E_{crit} = \Delta_{eff} \frac{(1 + 2\mu)}{(1 + \mu)}, \tag{5.3}$$

where the value of $\mu = m_c/m_v$ can be unity in the case of a symmetric band structure (one in which the Fermi level lies at the centre of the bandgap). However, $\mu < 1$ in the case of semiconductors [71]. Its value for water remains unknown; a value of

$\mu = 1$ is often used [34]. Following this assumption, we note that $E_{crit} = 1.5\,\Delta_{eff}$ for collisional ionization to occur in aqueous water. More detailed consideration of the energetics requires knowledge of the collision cross sections in water, and on their energy dependence [34, 72].

Quantitative insights into the temporal considerations that govern the ionization dynamics in water are also difficult to obtain but the following qualitative picture can be constructed subject to, as noted in the following, prevailing controversy. Avalanche ionization—ionization due to electron rescattering—has to proceed on slower time scales than direct optical field induced ionization. This is a consequence of the former requiring the occurrence of several IB events in cascade before E_{crit} can be attained by the quasi-free electrons. Given the 6.5 eV bandgap in water, the value of Δ_{eff} is about 7 eV for a Keldysh parameter of 2. Under these circumstances, collisional ionization by the quasi-free electrons would only occur when $1.5\Delta_{eff} \sim 11$ eV is attained. For initial optical field ionization by intense 800 nm light, this necessitates the ionized electron to undergo eight successive IB events before it can have enough energy to induce ionization. If we estimate the time between successive collisions in aqueous water to lie between 1 fs [73] and 2 fs [74], it requires about 13.6 fs for a doubling of the number of quasi-free electrons that can take part in inducing collisional ionization. This assumes that every collision involves absorption of a photon via the IB process. A detailed analysis of the temporal aspects [72] has led to the conclusion that avalanche ionization plays only a minor role in femtosecond laser induced breakdown of aqueous media like water. However, theoretical as well as experimental evidence is claimed which is supposed to illustrate exactly the opposite scenario: that avalanche ionization dominates the femtosecond laser induced breakdown of aqueous media like water [75–77]. The controversy awaits resolution!

5.4 Some Quantitative Considerations

To gain quantitative insights into the physics that drives cell surgery with femtosecond-long laser pulses, it is relatively simple to consider plasma formation in water that would occur in the vicinity of the threshold for optical breakdown—just below the breakdown threshold and just above it. But how does one determine the breakdown threshold? In the earlier days of laser surgery, the use of nanosecond and picosecond laser pulses produced breakdown that was inevitably accompanied by formation of a luminous plasma that could be readily seen by the naked eye and which served as a benchmark for the onset of optical breakdown. Cavitation and shock-wave emission also occurred, and observation of bubbles would also aid in estimating the incident fluence at which breakdown occurred. With the use of femtosecond pulses, plasma luminescence usually becomes visually obvious at fluence values that are much higher than those at the threshold for optical breakdown and recourse has to be taken to observation of bubble formation. However, as was subsequently realized, nanobubble formation can often occur that is not visually discernible although the effects of such bubbles can sometimes manifest themselves [78, 79]. These physical

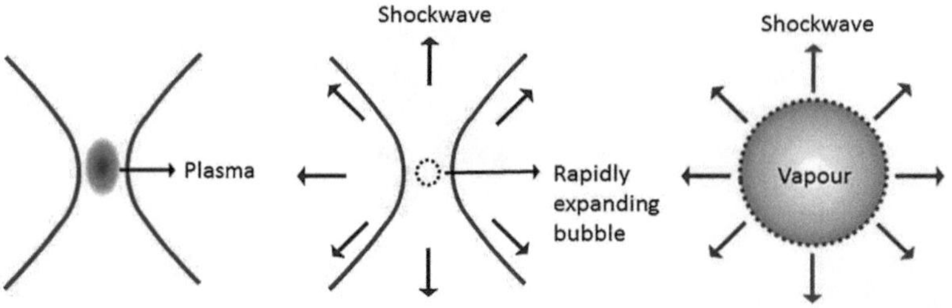

Fig. 5.2 Typical dynamics of the formation and "bursting" of nanobubbles, with resulting shockwaves which can have a major influence on laser surgery considerations

signals are usually due to effects of shock waves that are created upon the bursting of the nanobubbles (Fig. 5.2). Clearly, the disruptive nature of such shockwaves are of considerable relevance in the context of laser surgery.

In the absence of direct visual information pertaining to nanobubbles, recourse must to be taken to theoretical avenues to determine the optical threshold and, to this end, a critical electron density, σ_{cr}, within the laser focal volume is defined at which the plasma formed upon optical breakdown of water becomes strongly reflective and, at the same time, absorbing

$$\sigma_{cr} = \frac{\epsilon_0 m_c}{e^2}\omega^2, \tag{5.4}$$

where ϵ_0 is the vacuum dielectric permittivity. At incident laser wavelengths of 355, 512, and 1064 nm, the corresponding values of σ_{cr} are $\sim 9 \times 10^{21}\,\text{cm}^{-3}$, $\sim 4 \times 10^{21}\,\text{cm}^{-3}$, and $\sim 1 \times 10^{21}\,\text{cm}^{-3}$. A value of $\sigma_{cr} = \sim 1 \times 10^{21}\,\text{cm}^{-3}$ is commonly used to represent the breakdown threshold [34]. In most quantitative considerations it is assumed that the incident energies used in laser surgery are low enough to preclude consideration of nonlinear propagation effects, such as filamentation. It is also assumed that plasma formation occurs in pure water. It is, however, quite possible that proximate biomolecules and absorbing dyes may modify the energy level structure sufficiently to permit new linear as well as nonlinear absorption processes.

Numerical simulations can be carried out to determine how the ionized electron density (σ, the electron density in the conduction band) evolves with time upon exposure to intense laser pulses. A generic rate equation [61], as depicted in Fig. 5.3, is found to be of utility. Here, η^p denotes the ionization induced by the strong optical field—either via multiphoton ionization or tunneling—while η^c is the contribution to the electron density via the electron rescattering mechanism. This second contribution depends on, of course, the number of electrons that are already ionized, and this is denoted by σ^c. The last two terms in the equation (Fig. 5.3) represent losses that may occur due to (i) diffusion of ionized electrons from the laser focal volume and (ii) to electron-ion recombination. The diffusive losses also depend on σ^c; the losses due to recombination involve an electron-hole pair and, consequently, depend on σ^{2c}.

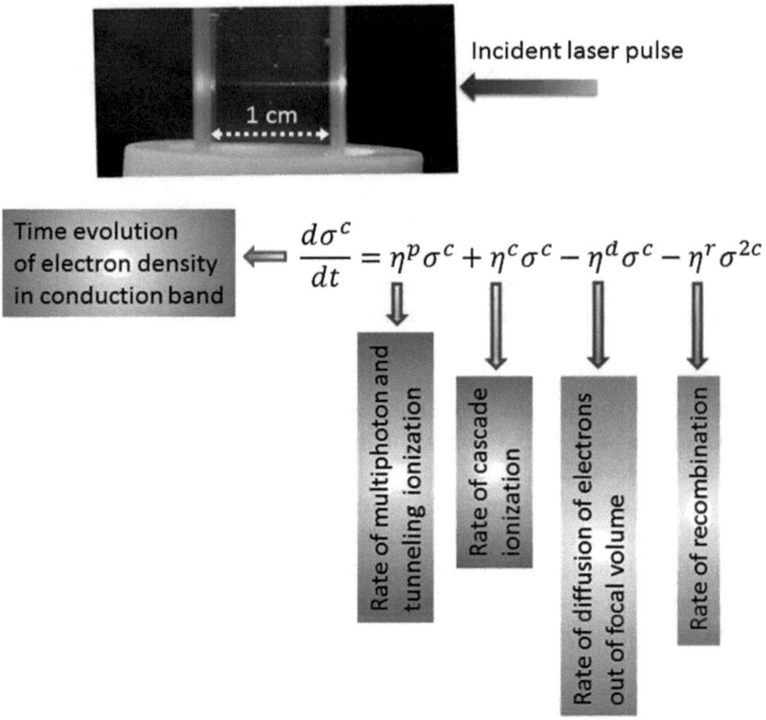

Fig. 5.3 The time evolution of the density (σ) of ionized electrons upon irradiation of biological matter with pulses of intense laser light. The *top* portion shows an intense, 40 fs laser pulse (800 nm wavelength) incident—from the *right*—on a 1 cm quartz cuvette containing water. Note the formation of a plasma channel (filament) within the water. The rate of change of σ with time is determined by the rate equation comprising the four terms that are shown. When the incident laser pulses are of femtosecond duration, the relatively slow loss mechanisms due to electron-ion recombination and diffusion out of the laser focal volume may be ignored

The loss terms shown in Fig. 5.3 account for electron loss processes that occur relatively slowly and they can safely be ignored in cases where ionization occurs as a result of irradiation by femtosecond pulses. However, they are important to consider when longer (nanosecond) pulses are used.

As has been demonstrated in comprehensive simulation studies carried out by Vogel and coworkers [61], it is recommended that instead of using the Keldysh parameter, γ that is of utility in gas-phase ionization of atoms (Chap. 2), it is often better (certainly more correct) to revert to the original formulation provided by Keldysh for photoionization in the condensed phase [80]:

$$\gamma = \frac{\omega}{e}\left(\frac{\epsilon_0 c_0 m \Delta'}{4I_e}\right)^{\frac{1}{2}}, \tag{5.5}$$

where I_e is the atomic ionization energy. The effective ionization energy, Δ', that is required in order to create an electron-hole pair in a condensed medium with band structure can be written as [80]:

$$\Delta' = \frac{2}{\pi}\Delta\frac{(1+\gamma^2)^{\frac{1}{2}}}{\gamma}G\left[(1+\gamma^2)^{-\frac{1}{2}}\right]. \tag{5.6}$$

Here, $G\left[(1+\gamma^2)^{-\frac{1}{2}}\right]$ is an elliptic integral of the second kind. Under conditions that $\gamma \ll 1$, the photoionization rate scales as I_e^k, where k denotes the number of photons required to jump the bandgap.

In computing the contribution made to the electron dynamics (Figs. 5.3 and 5.4), by the rescattering ionization term, it is often assumed that in the course of its rescattering trajectory, each participating electron gains an average energy from the

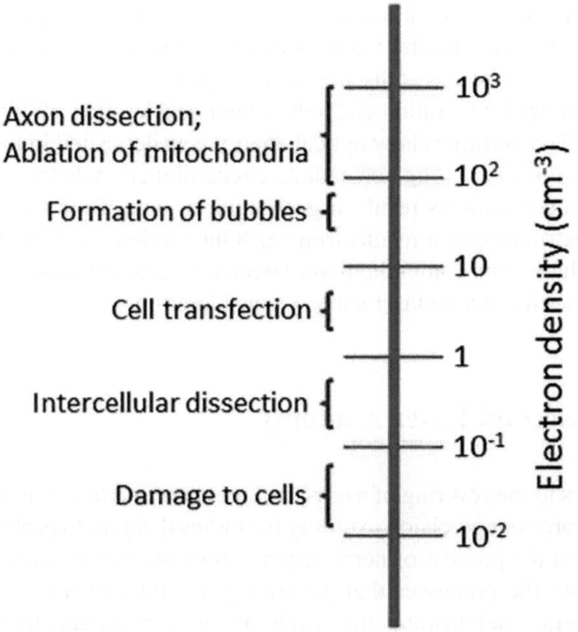

Fig. 5.4 An overview of some scales of "free" electron densities with reference to approximate thresholds for optically-induced processes like cellular damage, dissection, transfection, bubble formation and macroscopic damage like ablation. The threshold values assume that water is the medium in which the bio-sample is placed and that 800 nm light pulses of 100–200 fs duration are used. Such pulses are focused using large numerical aperture objectives (NA ~ 1.3–1.4). The largest electron densities ($\geq 10^{21}$ cm^{-3} require the use of an amplified laser system (usually operating at 1–10 kHz repetition rate) while the lower values of electron density are usually achieved with a femtosecond oscillator, typically operating at ~80 MHz repetition rate

optical field that is equivalent to $1.5\Delta'$ [34, 81]. Each such rescattered electron then contributes to the overall ionization rate as [81]:

$$\eta^c = \frac{1}{1+\omega^2\tau^2}\left[\frac{e^2\tau}{1.5\epsilon_0 c_0 n_0 m_c \Delta'} I - \frac{m_c\omega^2\tau}{M}\right], \tag{5.7}$$

where n_0 denotes the refractive index of the medium (usually water) at frequency ω, τ is the mean time between collisions, and m, M denote, respectively, the mass of the electron and the water molecule. At high values of incident laser intensity, I, the rate for rescattering ionization scales linearly with I. Values of τ have been estimated to be as fast as 1.7 fs [74]. It is also noteworthy that recombination of the ionized electrons with water molecules is not as simple a process as it would be in conventional gas-phase dissociative recombination. In the condensed phase that will inevitably be encountered in biophotonics, the ionized electrons become hydrated, on timescales of about 300 fs, and such hydrated electrons have a mean lifetime that has been reported to be as long as ~300 ns [82]. This is of relevance in the context of plasma-mediated damage by low-energy electrons and OH radicals that occurs upon irradiation of DNA in water by femtosecond-long pulses of intensity in the tens of TW cm^{-2} range [83, 84], as is discussed in Chap. 6.

The laser-induced generation and subsequent acceleration of “free” electrons in biological media determines how optical energy gets deposited in the medium. Collisional interactions, including non-radiative recombination, between such electrons and the plasma constituents result in generation of heat. Both the evolution of the temperature distribution that results from such interactions, and the subsequent bubble dynamics, have been comprehensively treated in a cogent review [34] in a manner that is of direct relevance to laser surgery considerations.

5.5 Femtosecond Laser Axotomy

Axotomy refers to the severing of axons, and femtosecond laser surgery has begun to be utilized to carry out precision axotomy in studies that aim to probe how genes and molecules affect the process of nerve regeneration and nerve development. Obtaining insights into the processes that govern regeneration of nerves is, of course, a critically important step towards the development of treatments for a host of neurological diseases [85]. Amongst the earliest realization that femtosecond laser surgery techniques (Fig. 5.4), by using high peak intensities, allow effective reduction of the energy threshold for removal of tissue (via ablation) and, thereby, allow a wide variety of surgery to be carried out with only an oscillator-based source, led to the use of femtosecond pulses to cut single axons inside the roundworm, *Caenorhabditis elangans* [47]. Pulse energies as low as a few tens of nJ in 200 fs long pulses of 800 nm light were shown to vapourize minute axon volumes (as small as 0.1 femtolitres, taking a mean axon diameter to be 300 nm). By filling the thus-axomotized neurons with an appropriate dye, it was possible to establish the physical disconnection of individual axons [47]. The low laser energies used in such experiments precluded the intervention of cavitation-induced physical effects and the 1 kHz repetition rate ensured that the average power was only about 10 μW, thus also precluding the possibility of thermal effects.

Larval-stage worms were used as laser surgery targets; circumferential axons located towards the dorsal side were laser-cut at their mid-body positions, with the rest of the axon being untouched. It was observed [47], via fluorescence imaging, that the laser-cut axons initially retracted. It was seen that axon regeneration occurred after about 24 h: the regenerated axons were fully functional, as indicated by the apparently normal locomotion of the worms.

References

1. A. Ashkin, J. Dziedzic, Science **235**, 1517 (1987)
2. S. Habuchi, R. Ando, P. Dedecker, W. Verheijen, H. Mizuno et al., Proc. Nat. Sci. Acad. USA **102**, 9511 (2005)
3. F. Zhang, L.-P. Wang, M. Brauner, J.F. Liewald, K. Ka et al., Nature **446**, 633 (2007)
4. A. Vogel, J. Noack, G. Huttmann, G. Paltauf, Appl. Phys. B **81**, 1015 (2005)
5. K. Svoboda, R. Yesuda, Neuron **50**, 823 (2006)
6. F. Helmchen, W. Denk, Nat. Methods **2**, 932 (2005)
7. P.J. Campagnola, L.M. Loew, Nat. Biotechnol. **21**, 1356 (2003)
8. D. Oron, D. Yelin, E. Tal, S. Raz, R. Fachima, Y. Silberberg, J. Struc. Biol. **147**, 3 (2004)
9. G.O. Clay, A.C. Millard, C.B. Schaffer, J. Aus-der-Au, P.S. Tsai et al., J. Opt. Soc. Am. B **23**, 932 (2006)
10. J.X. Cheng, X.S. Xie, J. Phys. Chem. B **108**, 827 (2004)
11. W.R. Zipfel, R.M. Williams, W.W. Webb, Nat. Biotechnol. **21**, 1369 (2003)
12. P.S. Tsai, P. Blinder, B.J. Migliori, J. Neev, Y. Jin et al., Curr. Opin. Biotechnol. **20**, 90 (2009)
13. G.D. Kymivonis, V.P. Kankariya, A.D. Plaha, D.Z. Reinstein, J. Refract. Surg. **28**, 912 (2012)
14. P. Kim, G.L. Sutton, D.S. Rootman, Curr. Opin. Opthalmol. **22**, 238 (2011)
15. S.I. Mian, R.M. Shtein, Curr. Opin. Othanlmol. **18**, 295 (2007)
16. P.K. Vaddavalli, S.H. Yoo, Curr. Opin. Othanlmol. **22**, 245 (2011)
17. C. Carrillo, A.S. Chayet, P.J. Dougherty, M. Montes, R. Magallanes et al., J. Refract. Surg. **21**, S655 (2005)
18. V. Venugopalan, A. Guerra, K. Nahen, A. Vogel, Phys. Rev. Lett. **88**, 078103 (2002)
19. K. König, I. Riemann, P. Fishcer, K. Halbhuber, Cell. Mol. Biol. **45**, 195 (199)
20. A. Vogel, W. Hentschel, J. Holzfuss, W. Lauterborn, Ophthalmology **93**, 1259 (1986)
21. M. Han, L. Zickler, G. Giese, M. Walter, F.H. Loesel, J.F. Bille, J. Biomed. Opt. **9**, 760 (2004)
22. T. Juhasz, F.H. Loesel, R.M. Kurtz, C. Hoverth, J.F. Bille, G. Mourou, IEEE J. Sel. Top. Quantum Electron. **5**, 902 (1999)
23. A. Histerkamp, T. Mamom, O. Kermani, W. Drommer, H. Welling et al., Graef. Arch. Clin. Exp. **241**, 511 (2003)
24. I. Ratkay-Traub, I.E. Ferincz, T. Juhasz, R.M. Kurtz, R.R. Krueger, J. Refract. Surg. **19**, 94 (2003)
25. W. Watanabe, T. Asano, K. Yamada, K. Itoh, J. Nishii, Opt. Lett. **28**, 2491 (2003)
26. W. Liu, O. Kosareva, I.S. Golubtsov, A. Iwasaki, A. Becker et al., Appl. Phys. B **76**, 215 (2003)
27. S.S. Mao, F. Quéré, S. Guizard, X. Mao, R.E. Russo et al., Appl. Phys. A **79**, 1695 (2004)
28. J. Kasparian, J. Solle, M. Richard, J.-P. Wolf, Appl. Phys. B **79**, 947 (2004)
29. M. Kolesik, E.M. Wright, J.V. Moloney, Phys. Rev. Lett. **92**, 253901 (2004)
30. U. Fuchs, U.D. Zeitner, A. Tünnermann, Opt. Exp. **13**, 3852 (2005)

31. C.B. Schaffer, A. Brodeur, J.F. García, E. Mazur, Opt. Lett. **26**, 93 (2001)
32. S. Tschachotin, Z. wiss Mikrosk. **29**, 188 (1912)
33. S. Tschachotin, Biol. Zbl. **32**, 623 (1912)
34. A. Vogel, J. Noack, G. Hüttmann, G. Paltauf, Appl. Phys. B **81**, 1015 (2005)
35. G. Hüttmann, J. Serbin, B. Radt, B. Lange, R. Birngruber, Proc. SPIE **4257**, 398 (2001)
36. D.G. Jay, T. Sakurai, Biochim. Biophys. Acta **1424**, M39 (1999)
37. S. Beck, T. Sakurai, B. Eustace, G. Beste, R. Schier et al., Proteomics **2**, 247 (2002)
38. G. Huettmann, R. Birngruber, IEEE J. Sel. Top. Quantum Electron. **5**, 954 (1999)
39. K. Schütze, G. Lahr, Nat. Biotechnol. **16**, 737 (1998)
40. K.R. Rau, A. Guerra, A. Vogel, V. Venugopalan, Appl. Phys. Lett. **84**, 2940 (2004)
41. K. Schütze, A. Clement-Sengewald, Nature **368**, 667 (1994)
42. S.K. Mohanty, M. Sharma, P.K. Gupta, Biotechnol. Lett. **25**, 895 (2003)
43. U.K. Tirlapur, K. König, Nature **418**, 290 (2002)
44. L. Paterson, B. Agate, M. Comrie, R. Ferguson, T.K. Lake et al., Opt. Exp. **13**, 595 (2005)
45. H. Schneckenburger, A. Hendiger, R. Sailer, W.S.L. Strauss, M. Schmitt, J. Biomed. Opt. **7**, 410 (2002)
46. W. Supatto, D. Dèbarre, B. Moulia, E. Brouzés, J.-L. Martin et al., Proc. Nat. Acad. Sci. USA **102**, 1047 (2005)
47. M.F. Yanik, H. Cinar, H.N. Cinar, A.D. Chisholm, Y. Jin, A. Ben-Yakar, Nature **432**, 822 (2004)
48. N.J. Smith, K. Fujita, T. Kaneko, K. Katoh, O. Nakamura et al., Appl. Phys. Lett. **79**, 1208 (2001)
49. S.W. Grill, J. Howard, E. Shäffer, E.H.K. Stelzer, A.A. Hyman, Science **301**, 518 (2003)
50. J.R. Aist, H. Liang, M.W. Berns, J. Cell. Sci. **104**, 1207 (1993)
51. A. Khodjakov, R.W. Cole, C.L. Rieder, Curr. Biol. **10**, 59 (2000)
52. H. Liang, W.H. Wright, S. Cheng, W. He, M.W. Berns, Exp. Cell Res. **204**, 110 (1993)
53. A. Heisterkamp, I.Z. Maxwell, E. Mazur, J.M. Underwood, J.A. Nickerson et al., Opt. Express **13**, 3690 (2005)
54. K. König, O. Krauss, I. Riemann, Opt. Express **10**, 171 (2002)
55. I. Riemann, T. Anhut, F. Stracke, R. Le Harzie, K. König, Proc. SPIE **5695**, 216 (2005)
56. W. Watanabe, N. Arakawa, S. Matsunaga, T. Higashi, K. Fukui et al., Opt. Express **12**, 4203 (2004)
57. L. Sacconi, I.M. Tolić-Norrelyke, R. Antolini, F.S. Pavone, J. Biomed. Opt. **10**, 014002 (2005)
58. K. König, I. Riemann, W. Fritsche, Opt. Lett. **26**, 819 (2001)
59. K. König, I. Riemann, P. Fischer, K. Halbhuber, Cell Mol. Biol. **45**, 195 (1999)
60. R.A. Meldrum, S.W. Botchway, C.W. Wharton, G.J. Hirst, EMBO Rep. **12**, 1144 (2003)
61. J. Noack, A. Vogel, IEEE J. Quantum Electron. **35**, 1156 (1999)
62. E.L. Botvinick, V. Venugopalan, J.V. Shah, L.H. Liaw, M. Berns, Biophys. J. **87**, 4203 (2004)
63. C.A. Sacchi, J. Opt. Soc. Am. B **8**, 337 (1991)
64. R. Lian, R.A. Crowell, I.A. Shkrob, J. Phys. Chem. A **109**, 1510 (2005)
65. P. Kambhampati, D.-H. Son, T.W. Kee, P.F. Barbara, J. Phys. Chem. A **106**, 2374 (2002)
66. D.M. Bartels, R.A. Crowell, J. Phys. Chem. A **104**, 3349 (2000)
67. M.U. Sander, M.S. Gudiksen, K. Luther, J. Troe, Chem. Phys. **258**, 257 (2000)
68. R. Lian, D.A. Oulianov, I.A. Shkrob, R.A. Crowell, Chem. Phys. Lett. **398**, 102 (2004)
69. S. Minardi, A. Gopal, M. Tatarakis, A. Couairon, G. Tamosauskas et al., Opt. Lett. **33**, 86 (2008)
70. D. Mathur, S.U. Khan, J.B. Hasted, J. Phys. B **11**, 3615 (1978)
71. B.K. Ridley, *Quantum Processes in Semiconductors* (Oxford University Press, Oxford, 1999)
72. B. Rethfeld, Phys. Rev. Lett. **92**, 187401 (2004)
73. N. Bloembergen, IEEE J. Quant. Electron. **10**, 375 (1974)
74. Q. Sun, H. Jiang, Y. Liu, Z. Wu, H. Yang, Q. Gong, Opt. Lett. **30**, 320 (2005)
75. A.C. Tien, S. Backus, H. Kapteyn, M. Murnane, G. Mourou, Phys. Rev. Lett. **82**, 3883 (1999)
76. A.P. Joglekar, H. Liu, G.J. Spooner, E. Meyhöfer, G. Mourou, A.L. Hunt, Appl. Phys. B **77**, 25 (2003)

77. A.P. Joglekar, H. Liu, E. Meyhöfer, G. Mourou, A.L. Hunt, Proc. Nat. Acad. Sci. USA **101**, 5856 (2004)
78. A.K. Dharmadhikari, J.A. Dharmadhikari, A.V. Mahulkar, G. Ramanandan, H. Ramachandran et al., J. Phys. Chem. C **115**, 6611 (2011)
79. H. Ramachandran, A.K. Dharmadhikari, K. Bambardekar, G. Ramanandan, J.A. Dharmadhikari et al., Nanotechnology **21**, 245102 (2010)
80. L.V. Keldysh, Sov. Phys. JETP **20**, 1307 (1965)
81. P.K. Kennedy, IEEE J. Quantum Electron. **31**, 2241 (1995)
82. D.N. Nikogosyan, A.A. Oraevsky, V. Rupasov, Chem. Phys. **77**, 131 (1983)
83. J.S. D′Souza, J.A. Dharmadhikari, A.K. Dharmadhikari, B.J. Rao, D. Mathur, Phys. Rev. Lett. **106**, 118101 (2011)
84. A.K. Dharmadhikari, H. Bharambe, J.A. Dharmadhikari, J.S. D′Souza, D Mathur, Phys. Rev. Lett. **112**, 138105 (2014)
85. P.J. Horner, F.H. Gage, Nature **407**, 963 (2000)

Chapter 6
Biophotonics in Ultrashort, Intense Optical Fields

Abstract Ultrashort pulses of intense laser light affect matter with which they interact and, in turn, matter also affects the light pulses. In this chapter we consider both "laser-matter" and "matter-laser" interactions from the viewpoint of biophotonic applications. Propagation of intense light through matter gives rise to visually spectacular phenomena like supercontinuum generation and filamentation. Such phenomena have been used to quantitatively probe damage induced in DNA by low-energy electrons and OH-radicals. They have also provided high sensitivity diagnostic of stress markers in human saliva. The chapter also discusses possibilities of biophotonics applications of such strong-field phenomena without the need for very high-intensity lasers.

6.1 Ultrashort, Intense Optical Fields

As has already been discussed in Chap. 2, intense optical fields are an inescapable consequence of ultrashort laser pulses produced in amplified laser systems. Ready availability of such lasers has now empowered researchers to carry out wide-ranging studies on how such ultrashort pulses of light propagate through transparent media, including media that is biologically relevant—such as water. Interest in such studies is propelled by fundamental considerations [1–3] and by tantalizing prospects of applications in diverse ventures like the control of lightening in the atmosphere [4–6] and remote sensing [7] on the one hand to the stand-off monitoring of atmospheric pollutants using broadband, time-resolved spectroscopy [8, 9] on the other.

When intense pulses of light induce multiphoton processes within materials with large bandgaps, energy is readily transferred from the incident optical field to the material that is being irradiated. Such energy transfer facilitates a number of practical opportunities in effecting bulk modifications in materials [10], such as forming—in controlled fashion—voids and micro-cracks due to micro-explosions and optical breakdown [11]. It is also readily possible to induce change of refractive index [12] which enable various processes to occur within materials, like the formation of colour centres and laser-induced oxidation-reduction of ions of transition metals, heavy metals and rare-earth elements [13]. Such laser-induced material modifications

P. Vasa and D. Mathur, *Ultrafast Biophotonics*, Biological and Medical Physics, Biomedical Engineering, DOI 10.1007/978-3-319-39614-9_6

prove to be of utility in the formation of waveguides [14] and photonic crystals [15], in the development of three-dimensional optical memories [16, 17] and, from a biophotonics perspective, in the formation of micro- and nano-fluidic channels *within* materials. Controlled laser machining within bulk material, without alteration of the surface properties of the irradiated material, is an area of considerable contemporary biophotonic interest and it can be easily appreciated that it is of utmost importance to develop proper insights into the major physical factors that govern the overall propagation dynamics [2, 3, 18].

There are two spectacular manifestations of the propagation dynamics involving ultrafast pulses of intense laser light: filamentation and white light generation (also known as supercontinuum generation). Both these processes coexist and serve as a signature that the laser-matter interaction is describable not merely in terms of laser-induced breakdown, as was discussed in Chap. 5. As shall become obvious in the following, the dynamics that accompany filamentation and supercontinuum generation are considerably richer than the formation of plasma due to breakdown. Filamentation manifests itself in the course of propagation through the medium while, at the same time, supercontinuum generation is visualized downstream of the medium, after propagation. Much remains to be accomplished before satisfactory insights are developed into the overall propagation dynamics in the strong-field, ultrashort regime. Nevertheless, it is eminently possible to *utilize* both filamentation and the supercontinuum in potentially useful biophotonic applications.

6.1.1 Filamentation

In the course of propagation an intense laser pulse experiences a refractive index that is intensity-dependent; this is a consequence of the optical Kerr effect (OKE) that was introduced in Chap. 2. OKE is the progenitor of the phenomenon known as "self-focusing", a process that occurs when the power of the incident laser pulse is greater than a certain critical power. The critical power for self-focusing, P_{cr}, is expressed in the continuous wave (cw) limit as [2]:

$$P_{cr} = \frac{3.72\lambda^2}{8\pi n_0 n_2}, \tag{6.1}$$

where n_0, n_2 are, respectively, the linear and nonlinear (intensity-dependent) refractive indices of the medium, and λ is the incident laser wavelength. For near infrared light ($\lambda \sim 800$ nm) propagating through air, the value of P_{cr} is $\sim$3 GW [19]. For propagation through condensed media of biological interest—water—the corresponding value is $\sim$4.3 MW.

The dynamics that drive the spatio-temporal wave collapse due to self-focusing was postulated by Townes and coworkers about half a century ago [20]. However, even at that time it was realized that the self-focusing action cannot continue indefinitely: it would eventually give rise to an infinity in intensity! Nature ensures that various mechanisms set in that help arrest the purely self-focusing action and, thus, avoid the infinity catastrophe. Such mechanisms include diffraction, group velocity dispersion, self-phase modulation, and pulse self-steepening, higher-order nonlinear effects, $\chi^{(5)}$ effects, and defocusing induced by plasma formation; all these contribute to the overall propagation dynamics in the form of single or multiple filaments or plasma channels. At very high values of incident laser power, multiple filamentation can occur which scales with laser power [21]. Factors affecting the length of filaments and the density of the plasma within these channels are the initial pulse intensity, initial pulse duration, and beam convergence [22, 23]. Note that, in the anomalous dispersion regime, it is expected that the self-guided laser pulses will spread less in time and space, affording long propagation lengths [24]. Consequently, theoretical as well as experimental efforts have recently begun to investigate filamentation and SC generation in the anomalous group velocity dispersion regime [25] and there might well be that important biophotonics implications emerge from such studies.

Figure 6.1 shows a typical image of a filament formed when 35 fs pulses (0.3 mJ energy) of 800 nm light are focused in water. Similar filaments are readily seen in a variety of condensed media, including crystalline solids, as long as the incident laser power exceeds the critical power for self-focusing (6.1). Visualization of such filaments is of utility, as shown in the following.

By imaging a filament, such as in Fig. 6.1, in the transverse direction it becomes possible to deduce the filament radius (L_{min}) as it propagates through the medium. Typically, the filament radius would be in the range 3–6 μm. From this, estimates can be made of peak intensities (I_{max}) as well as of electron densities (N_e) within the medium, by taking into consideration the Kerr effect, diffraction, and ionization responses. Following treatments detailed in [2, 3, 26, 27], the set of equations given below yield estimates of the peak intensity, electron density, and the multiphoton absorption cross-section, all from a simple determination of filament radius:

Fig. 6.1 Typical filament formed in water upon its irradiation by a focused beam of 35 fs long pulses of 800 nm light of incident energy 0.3 mJ. The incident laser light was linearly polarized

$$I_{max} \cong \frac{\rho_{max}}{2\rho_c n_0 n_2}, \tag{6.2}$$

$$\rho_{max} \cong \tau_p \rho_{nt} W(I_{max}), \tag{6.3}$$

and

$$L_{min} = 2\left(\frac{n_0}{k_0^2 n_2 I_{max}}\right)^{\frac{1}{2}}, \tag{6.4}$$

where τ_p is the pulse duration, ρ_c is the critical plasma density (see [28] for a discussion on filaments in a crystal of BaF_2), ρ_{nt} is the density of neutral atoms, and $W(I_{max})$ is the photoionization rate in the multiphoton (MPI) regime such that $W(I_{max}) = \sigma_k I_{max}^k$, where σ_k is the multiphoton absorption cross-section. The expression for L_{min} assumes a Gaussian beam profile and k_0 is a value related to geometrical properties of the focused laser beam [2].

6.1.2 White Light (Supercontinuum) Generation

The propagation of intense femtosecond-long laser pulses through matter also results in the formation, downstream of the region where filaments are formed, of a supercontinuum, a visually spectacular phenomenon that has continued to be interest for the past three decades [29, 30]. The continuum appears as a white disk that is surrounded by concentric, rainbow-like conical emissions (Fig. 6.2); the central, low-divergence part of the beam that is transmitted through the material has come to be known as the white light continuum or supercontinuum. The spectral width can be very broad, usually covering the range from the visible (~400 nm wavelength) to about 900 nm.

Various mechanisms have been invoked that might possibly rationalize the supercontinuum (SC) generation process: self-phase modulation [30], ionization-enhanced SPM [30, 31], with additional contributions from the interplay of diffraction and instantaneous electronic Kerr nonlinearity [32], stimulated Raman scattering, self-steepening, and four-wave parametric processes [33–36]. Notwithstanding the plethora of contributory mechanisms, a definitive understanding of the obviously rich physics that underpins SC generation continues to be elusive [37]. Pioneering experiments on the picosecond laser-generated supercontinuum [38] helped establish that the threshold power required for SC generation usually coincides with P_{cr}, the critical power for self-focusing. It, thus, appears to be the case that self-focusing also plays a role in generating the white light continuum [31, 39].

Although physical processes that describe SC generation are the same in all Kerr media, there remain a number of features that occur in water, a material of importance from a biophotonics perspective, whose true nature remain to be elucidated. Nevertheless, it is instructive to summarize some pertinent theoretical work on SC

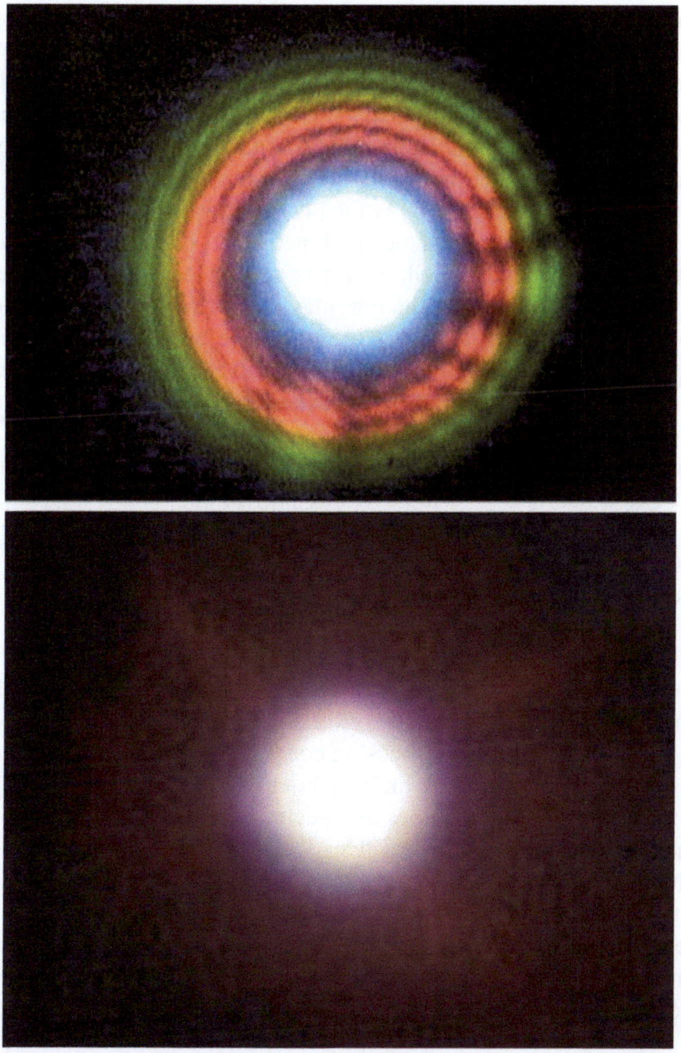

Fig. 6.2 Supercontinuum generation in water with 45 fs pulses of 800 nm light at different values of incident laser power. The image in the *upper panel* was measured at low values of incident laser power, ~50 P_{cr}, while the image in the *lower panel* was obtained at higher power, ~500 P_{cr}. The white central portion constitutes the supercontinuum and its spectral extent covers the range from ~400 nm to ~900 nm. The outer rings clearly visible in the upper panel are referred to as conical emission

generation specifically in water. A model [34] has proposed that the blue part of the SC spectrum—the region from ~400 nm to ~700 nm (Fig. 6.2)—occurs due to optical shock which results from the space-time focussing and self-steepening of the incident laser pulse. In this model, shock formation determines the dynamics of self-focusing, with multiphoton processes (resulting in plasma formation) helping stop the collapse. Theoretical simulations of propagation in water of 800 nm wavelength pulses of 144 fs duration and $\leq$20 μJ incident energy have yielded results that indicate that linear chromatic dispersion also contribute to the spectral width of the SC spectrum [40, 41], indicating that plasma generation helps determine the overall shape of the SC spectrum. Long-distance propagation of intense, ultrashort pulses in water is a process that is dynamic [42] and has been interpreted in terms of nonlinear X-waves that arise from the combined effect of diffraction, normal GVD, and self-focusing. As already noted, the robustness of long-distance pulse propagation in water is a consequence of the interplay between nonlinearity and chromatic dispersion. The dynamics of X-waves in water have been rationalized in terms of an effective three-wave mixing (TWM) model with SC generation occurring in the course of filamentation due to first-order scattering processes analogous to the first Born approximation [43].

In addition to material properties, the physical attributes of the light delivery system—the external optics—also come into play as do, indeed, the temporal characteristics of the laser pulse. Most early work on SC generation was conducted at incident power levels close to P_{cr}. Under such circumstances, rings and conical emission constituted a significant portion of the total output signal (Fig. 6.2). At higher levels of incident power, the central white light portion was seen to overwhelmingly dominate the total signal output [21]. By using loose focusing geometry and high incident power, SC efficiency levels in excess of ~40 % [21] have been demonstrated. Even such impressive conversion efficiency numbers can be further enhanced, by optimizing physical focusing conditions. Systematic studies have indicated that the position at which the incident laser beam is focused with respect to the irradiated material largely determines the spectral extent [44]. Such observations are a consequence of dispersion that broadens the pulse duration of the laser, thereby reducing its intensity: it has been shown in condensed media that the broadest white light spectrum is forthcoming when the external optics focuses the incident beam very close to the start of the medium, confirming that the dispersion does, indeed, play an important role. But, under real experimental conditions, it is often inadvisable to attain such focusing conditions as it frequently gives rise to extensive bubble formation when water is the irradiated medium [44].

The incident pulse duration, particularly at higher incident powers, also determines the spectral extent of the white light. Early experiments, conducted using 45 and 300 fs pulses at the same value of laser power [44], showed a significant difference in the spectral extent attained, especially in the blue region of the spectrum. The shorter pulses afforded a broader spectrum than 300 fs pulses, a consequence of the efficacy of SPM being inversely proportional to the pulse duration [29]. Free electron generation augments SPM, particularly leading to enhancement towards the blue spectral region [45]. Thus, using pulses of two different durations seems to indicate that the free electron contribution to SPM is larger in the case of 45 fs pulses than for longer (300 fs) pulses.

We now consider how SC generation depends on incident laser polarization and, also, the polarization properties of the supercontinuum itself. It has been demonstrated that a considerable improvement is obtained in the polarization extinction ratio of the SC generated via filamentation occurring in dual argon-filled gas tubes in tandem [46] as compared to the extinction ratio of the incident laser beam. Such improvement is most likely a consequence of significant improvement in the spatial mode quality of the laser beam brought about by filamentation. A consequence of the supercontinuum that is generated is, of course, the concomitant generation of ultrashort, few-cycle, laser pulses. It has been shown that it becomes trivially possible to control the plane of polarization of these few-cycle pulses by placing a half-wave plate *prior* to the dual-tube assembly [46]. A detailed discussion on polarization dependent SC generation, and the polarization properties of the SC that is generated, has been cogently discussed [46, 47]. In brief, the SC from water upon irradiation with 100 fs pulses has been shown to reduce when the incident light is circularly polarized [48]. Theoretically, it has long been held that the effective ionization energy for circularly polarized light is higher than for linearly polarized light [49] of the same intensity; correspondingly, MPI is a less efficient process when it is induced by circularly polarized light [50, 51]. In this context, it is worth noting that the value of P_{cr} for circularly polarized light is 1.5 times that for linear polarization [51]. The free electrons that are generated upon MPI augment SPM and contribute towards the generation of the blue side of the supercontinuum. The effect of light polarization on the multiphoton ionization of atoms has been related to the field correlations of multiphoton processes [52]. It is a consequence of the vectorial nature of the incident radiation field that affects the transition amplitudes of multiphoton processes: cross products of matrix elements involving the orthogonal components of the radiation field occur in the expression of the transition amplitude for circularly polarized light. These give rise to the dependence of ionization rate on the polarization state of the incident light field.

Filamentation- and supercontinuum-induced biophotonics generally relies on peak laser pulse intensities of the order of 10^{15} W cm^{-2}. At even higher pulse intensities, a new set of biophotonics opportunities open up that rely on laser-induced acceleration of charged particles—electrons as well as ions—which are beginning to be considered for use in applications like cancer therapy. But it is often the case that lower levels of incident power are beneficial for certain biophotonics applications. Are filamentation and SC generation of relevance at low power levels? The answer was discovered to be in the affirmative some years ago, as is described in the following.

6.1.3 Filamentation and Supercontinuum Generation at Low Power Levels

The use of intense light beams that are tightly focused often induce damage in the optical medium, and this can be a major concern from the viewpoint of biophotonics applications. It is, therefore, of interest to explore methods of broadband light generation *without* the need to focus the incident ultrashort laser beam with a lens. The methodology adopted for such experiments involves the use of a wire mesh and is based on the following premise.

It is expected that, in real life situations, the spatial profile of a train of femtosecond laser pulses will not have a perfectly smooth envelope of the type shown in Fig. 2.5 but will inevitably contain small amplitude irregularities. In the course of propagation through any medium such amplitude irregularities might break up into separate filaments. Such break-up has been described [53] in terms of interference between a strong beam and one or more weaker beams, giving rise to index variations—a consequence of the refractive index being intensity-dependent (the optical Kerr effect that we encountered in Chap. 2). A phase grating results from such interference, which may diffract the stronger light beam, thus enabling energy to be "spilt" into the weak beams. The filament formation process is, of course, stochastic and cannot be effectively controlled. Nevertheless, it has been shown [54] that a wire mesh placed in the path of the incident, unfocused laser beam radiation allows a certain measure of control to be exercised over the "spilling"of optical energy into the weak beams.

A wire mesh not only induces a modulated intensity distribution in the x-y plane but also in the direction of the laser beam propagation, the z-direction—the Talbot effect that is encountered in linear optics. Such intensity modulation is caused by interferences which produce local intensity maxima and minima. The maxima can serve as the seed for nonlinear processes in the optical medium and, as a consequence, multiple filamentation can occur due to the pre-existing (small-scale) inhomogeneities in the incident laser beam. Because of diffraction and interference, such inhomogeneities can be responsible for the creation of complicated "hot spots" along the z-direction. The diffraction induced by the wire mesh may be understood in terms of a 'normal' beam being affected by a stronger but controlled inhomogeneity pattern. If the mesh is regarded as a local focusing element, it becomes possible to determine the efficiency of off-axis (conical) emission. Schroeder and coworkers [54, 55] were able to demonstrate that conical emission can also be regarded as a loss channel that counterbalances catastrophic self-focusing discussed above.

The use of a wire mesh in the path of an unfocused laser beam implies that, in contrast to the usual situation that would be obtained if a focusing lens were used, the incident laser beam does not first converge and then diverge: the overall beam remains essentially parallel as it enters the medium to be irradiated. The wire meshes used in early experiments were made of stainless steel, typically of linear dimension 200 μm x 200 μm (or 440 μm x 440 μm) with wire thickness of 50–100 μm, so as to yield a typical 50 % transmission of light. By spatially scanning the mesh in a direction perpendicular to the laser propagation direction it was possible to monitor the amplitude irregularities expected due to the Talbot effect. Such scanning allows the irradiated sample experience different laser intensity profiles along the light propagation direction.

Figure 6.3 shows a typical energy profile of SC emissions from water upon its irradiation by a low-intensity, unfocused (parallel) laser beam of only 600 μJ energy. Altering the distance between the wire mesh and a 5 cm cuvette containing water along the direction of light propagation enabled water to be irradiated by different

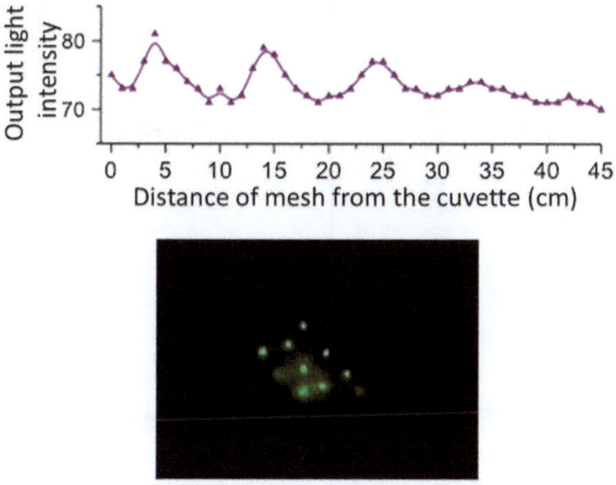

Fig. 6.3 White light generation in water (interaction length of 50 mm) with 45 fs pulses of very weak, *unfocused* 800 nm laser light. The incident pulse energy was only 600 μJ. The *top panel* shows prominent peaks and dips that are measured in the intensities of the white light emission in the far field (60 cm away from the sample). These reflect variation of laser energy generated at different distances between the mesh and water sample. The *solid lines* are a guide to the eye. The *lower image* shows multiple filaments that are formed in water

intensity profiles of the incident light, thus resulting in varying amounts of SC emission. Consequently, what is observed in Fig. 6.3 is a modulation of the intensity of the SC emission: positions of low intensity, which correspond to minimum or no white light, and high intensity (corresponding to maximum white light). Such fluctuations in the intensity of the SC emission have been simulated using Fresnel's diffraction integrals that enables calculations to be made of local intensities $[E_p(x_0, y_0, z)]^2$ at any point downstream of the wire mesh in terms of the optical field, E_p [55].

As already noted, the intensity maxima and minima seen in Fig. 6.3 reflect seeds for nonlinear processes in water. In that sense, the wire mesh can be thought of as simulating the "normal" case of laser beam propagation, where multiple filamentation can occur due to small-scale inhomogeneities in the incident beam. An example of multiple filamentation is shown in the image in Fig. 6.3. The processes of diffraction and interference can result in such small-scale inhomogeneities giving rise to complicated hot spot distributions also along the propagation axis.

It is also of importance to assess the extent of spectral broadening that is obtained using an unfocused laser beam. Figure 6.4 shows some typical spectra measured with a wire mesh, using very low values of incident laser energy. Marginal broadening is observed at 240 µJ but the extent of broadening is significantly enhanced when the energy levels are in the range beyond 400 µJ. Typical efficiency values for such white light emission were measured to be in the range of 30 % [55].

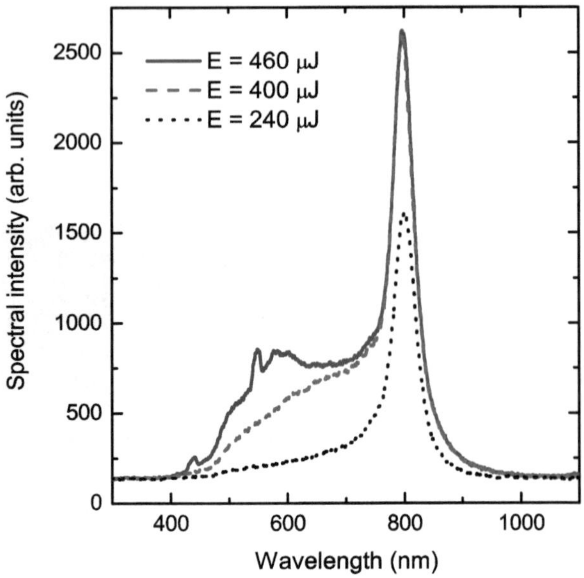

Fig. 6.4 Spectra of white light generated in water using 45 fs pulses of very weak, *unfocused* 800 nm laser light. The incident pulse energy was varied over the range 240–460 μJ by altering the distance between the wire mesh and the cuvette containing the water sample

6.2 Filamentation-Induced Biophotonics

6.2.1 DNA Damage

Recent years have seen the initiation of biophotonics experiments on aqueous DNA that are based on filamentation-induced dynamics. Irradiation of plasmid DNA in an aqueous medium by ultrashort pulses of intense laser light result in self-focusing and the formation of filaments within which optical field-induced ionization of water occurs. This gives rise to in situ generation of low-energy ionized electrons and free OH-radicals, the interaction of which with DNA plasmids has been shown to result in nicks in the DNA backbone: single strand breaks (SSBs) are induced as are, at higher laser intensities, double strand breaks (DSBs) which are not readily amenable to repair.

DNA is a naturally-occurring, long-chain polymer that comprises a number of nucleotide monomers linked to each other by phosphodiester bonds. Under physiological conditions, DNA is very stable: it has a half-life toward spontaneous hydrolysis of ~130,000 years [56]. It has high mechanical robustness, a consequence of its double stranded and double-helical structure. At the same time, nature dictates that, along with its mechanical stiffness, it must also be able to withstand fairly large conformational changes—like the bending, compression, and twisting of its 3D structure [57]—in order to be able to effectively pack into chromosomes. The ability of DNA to withstand large-scale structural change also serves it well in the primary task nature has decreed it to perform in the course of replication and transcription: the encoding and transferring of genetic information.

Exposure of this robust polymer to radiation causes damage which occurs upon the breaking of single and double DNA strands. Such damage can potentially be the most lethal to occur at the cellular level. Strand breakages occur when the sugar-phosphate backbone is ionized upon exposure of DNA-containing matter—that is, all living matter—to high energy radiation (such as energetic X-rays, γ-rays, and such like radiations from terrestrial and extra-terrestrial sources). Cosmic rays constitute a major extra-terrestrial source of energetic particles that interact with human cells; a plethora of chemical oxidizing agents are examples of terrestrial agents of DNA damage. Up to 500,000 damage events can occur per cell per day, but all of these are amenable to internal repair: the rate of damage in a healthy cell is more or less equal to the rate of repair. In a diseased cell, however, the rate of repair is less than the rate of damage, and this can lead to one of three main consequences:

(i) Senescence: an irreversible state of dormancy;
(ii) Apoptosis, or programmed cell death: this amounts to cell suicide;
(iii) Unregulated cell division: this can lead to formation of a tumour that is cancerous.

DNA defects can assume one of two main forms (Fig. 6.5): (a) either only a single strand has a defect, in which case the complementary strand can act as a template to biochemically correct the damaged strand, or (b) two strands are simultaneously damaged. This is particularly dangerous as double strand breaks may result in genome rearrangements within the cell.

6.2.2 *Strong Field Experiments with Intense, Ultrafast Laser Pulses*

About a decade ago electron collision experiments on DNA (in dry form) established a new paradigm: damage to DNA is not only caused by energetic radiation. Even

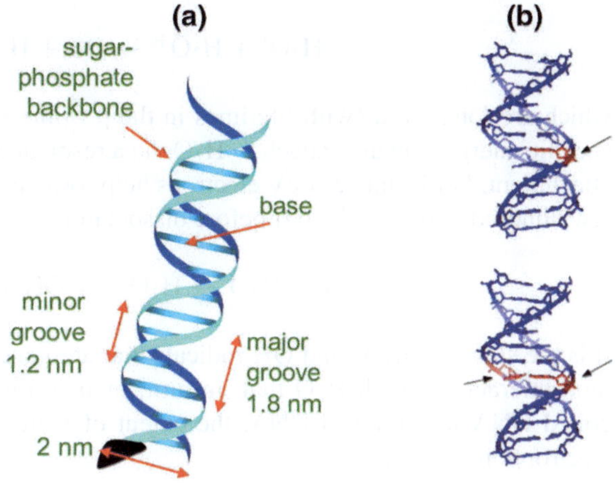

Fig. 6.5 **a** Schematic structure of the double-helix structure of DNA. **b** Schematic depiction of damage caused to DNA in the form of a single strand break (*arrow* in the *upper figure*) or a double strand break (*two arrows* in the *lower figure*)

slow electrons, possessing only a few electron volts of kinetic energy, can induce strand breakages via formation of dissociative temporary negative ion states [58]. Subsequent experiments reported damage to DNA in its native, *aqueous* state by both slow electrons and radicals generated, in situ, upon irradiation of water (in which plasmid DNA was suspended) by 45 fs pulses of intense (1-12 TW cm^{-2}) 820 nm laser light [59]. These low-energy particles—both electrons as well as OH radicals—induced transformation of supercoiled DNA into relaxed DNA, a signature of strand breakages (Fig. 6.6). Subsequently, intense field experiments were conducted using laser pulses of longer wavelengths, 1350 and 2200 nm, at the same intensity levels as in the case of 820 nm light (in the TW cm^{-2} range), and the results of these served to delineate the role played by OH-radicals in inducing strand breaks in DNA under physiological conditions [60].

Biologists have long known that enzymes—like gyrases and topoisomerases—exhibit a propensity for inducing nicks in the DNA backbone. The observation that nicks may also be produced in the course of interactions of aqueous DNA with intense optical fields strongly suggests that the latter are able to mimic the nick-inducing action of enzymes. This has biophotonics implications beyond studies of DNA per se as it highlights the ability of the filamentation-based method of generating slow electrons and radicals within aqueous media as being of importance in diverse situations in which the effects of low energy radiation need to be probed under physiologically relevant conditions. There is also peripheral interest in that high-intensity laser light of wavelength longer than 1300 nm is presently characterized as eye-safe in industry. The observation that strong irradiation of water at such wavelengths induces strand breakages in DNA raises the obvious question: how safe is "eye-safe"?

In view of the relative novelty of filemantation-based biophotonics, it is of interest to present some experimental details. In experiments on aqueous DNA the key initiator of the damage-inducing dynamics is the strong optical field that initiates the process of excitation, ionization, and dissociation of H_2O molecules, yielding species like $H_2O^\star$, H_2O^+, OH, $OH^\star$, and slow electrons. Collisions between electronically excited states of water and ionized water molecules give rise to the formation of OH radicals [61]

$$H_2O^* + H_2O^+ \rightarrow OH + H_3O^+, \tag{6.5}$$

which are long-lived (with lifetimes in the μs time scale) [62]. Slow electrons, of specific energy, can also attach to H_2O via a resonant process known as dissociative attachment. For instance, 7 eV electrons help form an H_2O^- state that survives for a few hundred attoseconds [63] before dissociating:

$$e + H_2O \rightarrow H_2O^- \rightarrow OH + H^-. \tag{6.6}$$

It is the slow electrons and OH-radicals that are generated, in situ, in these strong-field interactions with H_2O that, in turn, induce transformation of initially supercoiled DNA into relaxed DNA, the extent of which can be quantified using gel electrophoresis [59].

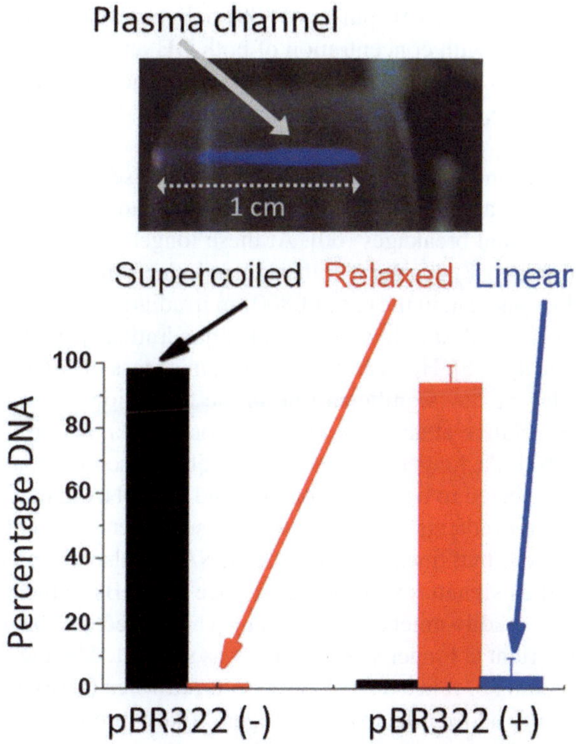

Fig. 6.6 Irradiation of water containing DNA plasmids (pBR322) by intense, 40 fs long laser pulses induces the formation of a plasma channel (filamentation), as depicted by the bluish channel shown in the *upper image*. After irradiation for 30 s, the DNA is separated—offline—using gel electrophoresis. Post-separation, the gel is stained with ethidium bromide, a DNA binding fluorescent dye, that helps quantify the amounts of DNA in different forms (see text). The *lower panel* shows that prior to laser irradiation [pBR322 (−)], almost all (∼97 %) of DNA is in the supercoiled state, with only about 3 % in the relaxed stated. After 30 s of laser irradiation, [pBR322 (+)], the ratios are almost reversed: now ∼95 % of DNA is in the relaxed state while only about 2 % is in the supercoiled state. There is also a component that is DNA in linear form. Relaxed DNA structures are indicative of single strand breaks; linear DNA is a signature of double strand breaks

Typically, only a tiny fraction of DNA suspended in water might be expected to lie within the laser-induced plasma channel (Fig. 6.1). But it has been established [64, 65] that strong thermal gradients are set up as the laser beam propagates through water. These gradients set up convective flow such that DNA molecules within the interaction volume are constantly replenished.

The question that arises from the results shown in Fig. 6.6 is: "Is the damage to DNA caused by low-energy electrons or by low-energy OH-radicals?"

A technique that has been effectively utilized [59, 60] to delineate the effects of electrons and OH radicals relies on the introduction into DNA-doped water of minute quantities of chemical scavengers of low-energy electrons and/or OH radicals. Of course, such chemicals have to be judiciously selected such that they do not, on their own, induce single strand breaks (SSBs) and/or double strand breaks (DSBs).

Experiments conducted with 40 fs pulses of 800 nm light revealed that the percentage of relaxed DNA varied with concentration of both OH and electron scavengers; the quenching rate that was measured by varying the scavenger concentrations indicated that OH radicals were, in fact, four times more likely to cause SSBs than slow electrons [59].

Subsequent experiments made use of ultrashort pulses of longer wavelengths –1350 and 2200 nm—at intensities in the TW cm^{-2} range, to further clarify the role of OH-radicals in strand breakages [60]. At these longer wavelengths the ponderomotive potential (Chap. 2) accelerates the ionized electrons (prior to solvation) to much higher velocities than in the case of 800 nm irradiation. This is simply because the accelerated electrons' kinetic energy scales quadratically with wavelength and linearly with intensity [66]. Higher energies (say, in excess of ~10 eV) preclude the electrons from playing any significant role in the formation of temporary negative ions states as dissociative attachment cross sections are known to fall rapidly with electron energy [67]. At longer-wavelength, therefore, the electron contribution to DNA damage is expected to be "switched off" and any DNA damage is likely to be solely induced by OH-radicals. In the course of these longer-wavelength experiments [60] it was discovered that linear structures of DNA are also detected (Fig. 6.6) and these are regarded as signatures of the occurrence of DSBs (which, as has already been noted, are not readily amenable to repair). The conjecture that electron effects are no longer important at longer wavelengths was supported by the observation that pronounced quenching of relaxed DNA only occurred when OH-scavengers (such as sodium acetate) were added; electron scavengers (like 5-bromouracil) had no effect [60]. This allowed the deduction to be made that the *supercoiled* → *relaxed* transition observed when irradiation is by longer wavelength pulses is exclusively induced by OH-radicals [60].

Experiments with OH scavengers also enabled consideration of the possible role of thermal effects at these long wavelengths. It was shown [60] that with 1350 nm photons and high scavenging efficiency values ($500 \times 10^{12}\,s^{-1}$), SSBs are quenched by ~90 % while with 2200 nm photons they quench to ~80 %. In other words, there is more thermally-induced damage at 2200 nm than at 1350 nm photons, in line with higher absorption by water of 2200 nm light. Thermal gradients are, thus, not a major contributor to strand breaks although, as noted above, they contribute to the effective "stirring" of the DNA plasmids throughout the cuvette containing water.

What of damage being induced due to conventional photoinduced effects? The photoabsorption spectrum of DNA indicates that maximum linear absorption occurs in the case of 260 nm photons, and that such photoexcitation is a likely cause of lesions, including DSBs [68]. Photoexcitation with 1350 nm and 2200 nm wavelength photons would necessitate 6-photon and 9-photon excitation, respectively. At a fixed value of photon flux, multiphoton induced damage to DNA occurs more readily in the case of 1350 nm photons than with 2200 nm photons. However, the results of long wavelength experiments indicated that at intensities of the order of 25 TW cm^{-2}, DSBs are formed only with 2200 nm photons and not with 1350 nm photons [60], seemingly clear-cut proof against multiphoton effects playing a role in the DNA damage dynamics.

6.2.3 Strand Breakages: Possible Mechanisms

If the experimental evidence is so overwhelming that it is OH radicals that are the cause of SSBs and, more importantly, DSBs, it is pertinent to probe how these strand breaks are induced. It is known from experiments with high energy X-rays and γ-rays that collisions of OH with DNA account for the majority of radiation damage to cellular systems [69]. The precursor to the damage is collisionally induced H-atom abstraction from one of the five unique carbons of the deoxyribose pentose ring. Collisions of OH radicals with sugar H-atoms give rise to SSBs although it is still not established whether such breaks occur directly or via some indirect route [70, 71]. By direct collision is meant an OH-induced DSB occurring via radical transfer between complementary DNA strands [72]. An indirect collision, on the other hand, implies that energy is inhomogeneously deposited in the host H_2O molecules such that clusters of OH radicals are formed: each DSB then occurs when two proximate SSBs are induced by separate OH radicals from within such a cluster. This corresponds to a "double-collision" process. The scavenger experiments conducted at longer wavelengths shed some further light on these possible scenarios: DSBs were found to be scavenged much more readily than SSBs. In other words, the probability for a two-collision event reduced much faster than a direct (single-collision) event, providing indications in favour of an indirect "two-collision" process.

Systematic experiments that have been carried out to quantify SSBs and DSBs in DNA plasmids, with different values of incident laser intensity at 820, 1350, and 2200 nm wavelengths, have lead to the conclusion that DSBs are induced *primarily* by wavelength effects rather than by laser intensity effects. In this connection we note that, at an intensity of 25 TW cm^{-2}, values of the ponderomotive potential, U_p, which the ionized electrons experience are 1.5 eV at 820 nm, 4 eV at 1350 nm, and 11 eV at 2200 nm.

6.2.4 Role of Water Molecules: Electronic Structure Considerations

Electron collisions with H_2O molecules will lead to the following scenario vis-a-vis electronic excitation of water: at electron energies equal to, or in excess of, 9.13 eV the electronically excited $H_2O^\star$ state can undergo dissociation. In this case, the dissociation process is adiabatically correlated to OH fragments formed in their electronically excited $A^2\Sigma^+$ state. The lowest energy electronic configuration of H_2O is $(1a_1)^2(2a_1)^2(1b_2)^2(3a_1)^2(1b_1)^2$, yielding a ground electronic state of X^1A_1 symmetry. Excitation of an electron from the $1b_1$ to the $4a_1$ orbital requires an energy of $\sim$7.3 eV, producing the electronically excited $H_2O^\star$ state, A^1B_1, which dissociates to OH in its ground electronic state, $^2\Pi$. With only 1.6 eV more energy, a $3a_1 \rightarrow 3sa_1$ excitation can be induced. This leads to formation of a higher-lying $H_2O^\star$ state, B^1A_1, which can adiabatically dissociate to excited $OH^\star$ $(A^1\Sigma^+)$. However, dominantly, it nonadiabatically dissociates to produce rotationally hot OH ($^2\Pi$) fragments [73]. Such rotationally excited OH fragments have as much as 4 eV of rotational energy! The high rotational angular momentum in these OH fragments is a consequence of the torque that is generated in the proximity of a conical intersection between potential energy surfaces of the B and X states in the collinear H-O-H geometry [73]. Rotationally hot OH fragments may have a high propensity (high cross section) to abstract sugar H-atoms via simultaneous collisions at two different sites, giving rise to DSBs. This scenario was tested [60] in experiments conducted using 1350 nm photons but at a high enough intensity (75 TW cm^{-2}) to ensure a U_p value of 11 eV. With such energetic electrons, about ~4 % of the irradiated DNA did, indeed, appear in linear form.

Evidence seems to strongly suggest that electronic excitation of H_2O such that dissociation of the resulting $H_2O^\star$ produces rotationally excited OH fragments is a critical facet of the overall dynamics that lead to formation of DSBs. At very low intensities (5 TW cm^{-2}) of 2200 nm light, there is no evidence of linear DNA being formed [60]. This is because the energy to which ionized electrons are accelerated by U_p is insufficient to electronically excite H_2O molecules to states that dissociate into hot OH fragments. However, upon increasing the intensity to ~30 TW cm^{-2}, as much as 20 % of the irradiated DNA is found to be in linear form. It is, therefore, likely to be the case that attosecond duration dynamics involving ionized electrons mediate DNA damage dynamics in these strong field experiments!

6.3 Supercontinuum Biophotonics

6.3.1 Supercontinuum-Based Stress Diagnostics in Humans

Propagation of ultrashort, intense laser pulses through medium that contains biological matter has considerable potential in biophotonics, specifically in biomedical applications. Early studies probed the propagation of ultrashort (<45 fs) pulses of 820 nm light through water doped with minute quantities of biological matter, particularly tiny concentrations of salivary proteins like α-amylase and various amino acids [74]. It was observed that with α-amylase dopants in water, there was a very significant concentration-dependent suppression of SC generation. The interest in α-amylase in these pioneering studies stemmed from its importance as a potential marker of stress experienced by humans. There was biochemical evidence that indicated a positive correlation between physiological and psychological stress and changes brought about in α-amylase activity in human saliva.

A direct relationship seems to have been established between salivary α-amylase and biochemical stress markers like catecholamines [75]. This is of importance as α-amylase has essentially no chemico-physiological relationship with catecholamine molecules or, indeed, with other biochemical stress markers like cortisol. Hence, α-amylase that is found in saliva might well turn out to be an additional, independent parameter in measurements of stress in humans. As a potential stress marker, α-amylase offers a distinct and important advantage of being non-invasive, readily and cheaply obtainable, and sensitive [76]. Results of microarray experiments have offered indications of enhanced α-amylase production in microbial cultures in the presence of oxidative stress [77]. Oxidative stress is usually a consequence of the presence of reactive atomic oxygen in a culture. Attempts have been made to develop sensors based on α-amylase activity to quantify stress. Even in plant systems, although water-induced stress does not cause a significant change in total protein synthesis it does, nevertheless, give rise to an enhancement in α-amylase mRNA, indicating that regulation of gene expression might well result from metabolic changes brought about by water-induced stress [78].

It is pertinent to make a few remarks on salivary proteins other than α-amylase. It now appears established that saliva contains a number of chemical markers that are of interest in the context of screening, diagnosing, and monitoring clinically-significant information [79, 80]. Collection of saliva is easy, usually involving gentle rubbing of a cotton swab on the inner cheek or by spitting into a tube. Significant salivary biomarkers consist of biomolecules like the globulins (immunoglobulins, for instance), enzymes (like α-amylase and lysozyme), other sundry proteins (like statherin and mucin), hormones (like cortisol), toxins and their metabolites (cotinine), DNA, and traces of various heavy metals (such as lead) [80]. Thus, human saliva is a rich repository that can be used as a clinical specimen with potential to yield biomedical and physiological information in inexpensive and efficient manner. It also appears that saliva can yield information that cannot be readily extracted from blood serum. Hence, there are sound reasons that saliva-based diagnostics have begun to attract the attention of the biomedical community [79–81].

Globulins, which comprise α- and β-globulins, are present in human blood; their concentration is second only to albumins. Immunoglobulins fall in the group of α-globulins and are antibodies that are produced in blood as it flows in the human circulatory system in order to offer protection from "invasion" of disease-producing organisms [82].

Saliva-based studies have suggested that the concentration of salivary constituents other than α-amylase, like immunoglobulin A (IgA), can also be correlated with psychological stress which affects certain oral pathology conditions [83–87]. For instance, studies have offered indications that the IgA level in human saliva is significantly higher on the day before academic examinations and during them; the level lowers on the days between these examinations. This opens the possibility of using the level of IgA as a promising technique to monitor psychological stress in students, workers [87] and, presumably, authors of books as the publisher's deadline looms large. Some correlation also seems to have been established between a person's sense of humour and her/his secretory IgA levels [88]!

Experiments to measure SC generation in water and water doped with minute quantities of salivary proteins have been conducted with 45 fs long pulses of 820 nm light from a Ti-sapphire laser operating at 1 kHz repetition rate at typical incident energies corresponding to $\sim$100–400 P_{cr}. A typical SC spectrum (Fig. 6.7) indicates the presence of two components:

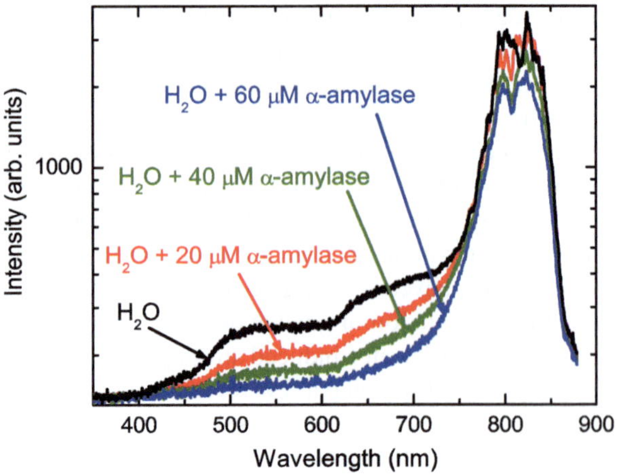

Fig. 6.7 Spectra of white light generated in pure water as well as water doped with minute concentrations of a salivary protein, α-amylase. 45 fs pulses of focused 820 nm light were used to attain an incident intensity of 2 TW cm^{-2}

(i) Around 820 nm there is a contribution from self-phase modulation (SPM) that arises from the Kerr nonlinearity; as expected from the discussion in Sect. 6.1, the resulting spectral broadening is essentially symmetric about 820 nm.
(ii) There is also a second component which is asymmetric, towards the blue side. This arises from processes like space-time focusing, self steepening, and plasma formation that results when free electrons are generated upon field-induced ionization of water. These electrons significantly affect the white light spectrum in terms of a positive frequency deviation $\delta\omega(\tau_p)$ that results from the negative contribution to the refractive index Δn from the ionized electrons. As already discussed, the extent of asymmetric spectral broadening that is observed in Fig. 6.7 can be modelled by considering contributions made to the refractive index by the Kerr nonlinearity, Δn_K, and by the plasma, Δn_P [89]:

$$\delta\omega(\tau_p) = -kz\frac{(\Delta n_K + \Delta n_P)}{\tau_p}, \tag{6.7}$$

where z denotes the propagation distance and k is the wave number.

In addition to asymmetric broadening (Fig. 6.7), it is also noticeable that there is a significant dip at ~625 nm that is superimposed on the supercontinuum. This is due to an inverse Raman effect [90]. Upon irradiation of water by intense monochromatic light of frequency ν_0 and, simultaneously, also with intense white light, the H_2O molecules within the laser focal volume are stimulated: they emit radiation at frequency ν_0 while concomitantly absorbing radiation at frequencies $(\nu_0 + \nu_m)$ and $(\nu_0 - \nu_m)$, where $h\nu_m$ is the energy separation between different vibrational levels in a given electronic state of H_2O or H_2O^+. It is absorption on the higher frequency side that accounts for the dip seen in Fig. 6.7; it corresponds to the well-known 3500 cm^{-1} Raman shift in water [90].

The most noticeable feature of Fig. 6.7 is the change in the shape of the spectrum as very small quantities of α-amylase are added to water. Even at very small molar concentrations, like 20 μM, the extent of spectral broadening markedly decreases. In these experiments the incident laser intensity was kept constant, so the extinction of the white light—as manifested in the decreased broadening of the SC spectrum—is purely attributable to α-amylase.The extent to which SC production is curtailed by the salivary protein is also reflected in corresponding extinction of the inverse Raman spectral feature. At 60 μM concentration the remnant broadening is seen to be more or less symmetrical around the incident laser wavelength: it represents SC production from self-phase modulation effects. This symmetric broadening hence indicates the extinguishing of plasma-induced contributions to SC production. α-amylase extinguishes the plasma by effectively scavenging the free electrons that are formed in the course of optical field-induced ionization of water [74]. The disappearance of the free electrons is what leads to suppression of plasma-induced (asymmetric) spectral broadening and leaves behind the remnant SPM-induced supercontinuum. This "disappearance" of electrons is attributed to the dissociative attachment process which, for a triatomic molecule ABC, can be generically represented as

$$e + ABC \rightarrow ABC^{-} \rightarrow AB^{-} + C, \tag{6.8}$$

or

$$e + ABC \rightarrow ABC^{-} \rightarrow AB + C^{-}. \tag{6.9}$$

The important consequence of such dissociative attachment (DA) collisions is that while the reactant part of the equation contains a free electron, the product side does not. DA, therefore, leads to the scavenging of free electrons. The DA process is a resonant process and, typically, DA cross sections are very large, sometimes as large as 10^{-14} cm^{-2}, for low energy ($\leq$10 eV) electrons. The temporary negative ion that is the intermediate step in the DA process is usually very short-lived. In the case of water, the resonant H_2O^- state in the vicinity of 7 eV survives for only a few hundred attoseconds [63].

The effective scavenging of electrons extinguishes the plasma, leading to SC generation that is only SPM-driven and, consequently, symmetric around the incident laser wavelength.

Is the suppression of the white light a property that is specific to α-amylase? More than 300 different proteins have been detected in human saliva, and there may be many more. Unsurprisingly, their individual functions are not yet established although it appears that only α-amylase and IgA amongst them are linked to physiological and psychological stress in humans.

A number of proteins found in blood serum have also been investigated from the perspective of SC generation [91, 92] and typical results for two such proteins are shown in Fig. 6.8. Note that this figure indicates a measure of SC suppression in the case of IgA, the protein found in saliva but there is no such suppression evident in the case of a non-salivary protein, IgG. The latter is found in the human pancreas. Even relatively large concentrations, up to 8 % by volume, of IgG failed to indicate

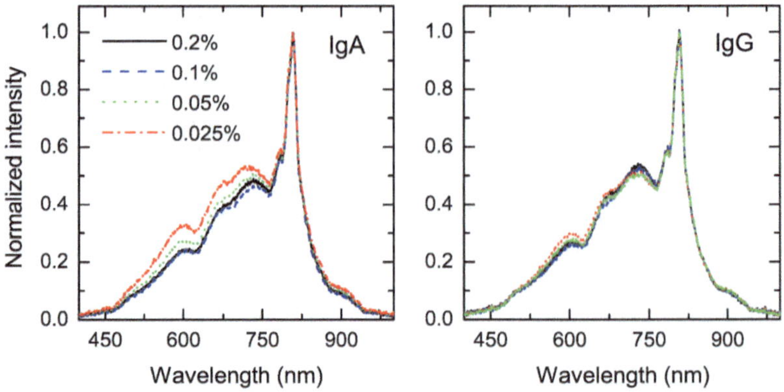

Fig. 6.8 Spectra of white light generated in water mixed with (*left panel*) two different doping levels of the salivary protein immunoglobulin A (*IgA*) and (*right panel*) the non-salivary protein immunoglobulin G (*IgG*), extracted from blood serum. IgG doping levels as high as 1 % were used. 45 fs pulses of focused 820 nm light were used to attain an incident intensity of 2 TW cm^{-2}

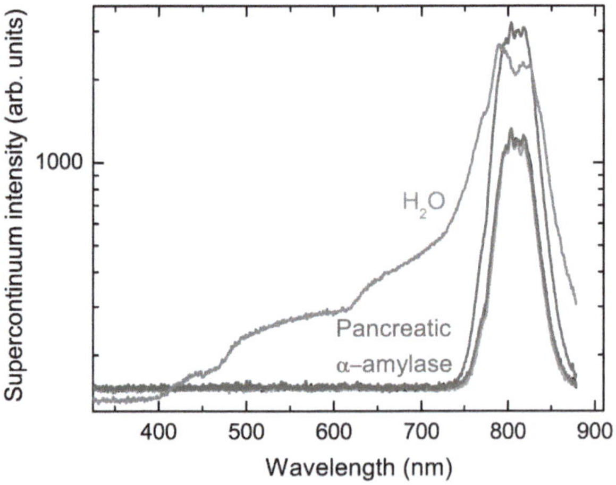

Fig. 6.9 Spectra of white light generated in water doped with pancreatic amylase. 45 fs pulses of focused 820 nm light were used to attain an incident intensity of 2 TW cm^{-2}. Note the *total* suppression of white light

SC suppression. Similar results have been reported in the case of other pancreatic proteins, like transferrin and human serum albumin [91]. In contrast, SC generation in amylase extracted from human pancreas has shown an extraordinary efficacy for *total* suppression of white light (Fig. 6.9).

The biomedical community knows that human saliva contains a plethora of electrolytes, immunoglobulins, proteins, enzymes, mucins, and nitrogenous products. Salivary amylase serves the purpose of catalyzing the chemical breakup of polysaccharides (starch) into smaller molecular entities and, in doing so, producing maltose as a product. Human salivary amylase can be found in two different forms [93]: a glycosylated isoform (of mass around 62 kDa) and a non-glycosylated version (of mass around 56 kDa) consisting of just under 500 amino acids with three domains.

The results shown in Fig. 6.7 correspond to an admixture of both these forms. Further experiments are needed to quantify their ratio. Salivary amylase content in human saliva becomes elevated in diverse situations, such as ruptured tubal malignancy, ovarian malignancy, lesions on the salivary glands, under conditions of renal insufficiency, acute misuse of alcohol, diabetic ketoacidosis, macroamylasemia, septic shock, cardiac surgery, tumours, and a variety of drugs [94]. Pancreatic amylase activity is related to pancreatic trauma, biliary tract diseases, intestinal obstructions, perforated peptic ulcers, gastritis, duodentitis, acute appendicitis, peritonitis and trauma [94]. Santhosh and coworkers [91] have made the following observation that seems pertinent to the present discussion. Only salivary proteins seem to exhibit the property of being able to effectively scavenge low-energy electrons appears to be consistent with saliva's anti-oxidant properties. For humans, saliva presents the first line of defence against oxidative stress induced by an onslaught of free radicals [95–97] and the electron scavenging properties that are being ascribed to salivary proteins are not inconsistent with their already established free-radical scavenging (anti-oxidant) properties. Non-salivary proteins found within the human body, on the other hand, have no need to perform electron scavenging tasks!

From a purely optics viewpoint, it may be possible to postulate that the shape of the SC spectrum alters upon addition of protein dopants because of induced changes in the chromatic dispersion. However, as has been pointed out [91], this postulate begs the question as to why only the salivary proteins and not the pancreatic proteins induce alterations in chromatic dispersion in water.

6.3.2 Supercontinuum Generation in Water at Long Wavelengths

The last few years have seen increased availability of femtosecond laser sources that emit at longer wavelengths than the usual 800 nm and this has enabled a new generation of theoretical and experimental investigations into filamentation and SC generation in the anomalous group velocity dispersion (AGVD) regime [89, 98–103]. In the AGVD regime plasma channels that are formed as a result of focusing-defocusing-refocusing cycles are expected to not spread (in time and space) as much as they would in the normal GVD regime [24]. Under normal GVD conditions, the collapse of the laser beam due to self-focusing occurs along the transverse spatial dimensions. In contrast, under AGVD conditions, beam collapse occurs in both the spatial as well as the temporal dimensions; the latter manifests itself as additional collapse in the longitudinal direction. As already discussed, such collapse is arrested when a plasma is formed: the free electrons in the plasma provide a defocusing mechanism by lowering the refractive index and enabling the pulse energy to be absorbed by the plasma. Thus, in case of normal GVD, optical energy is rapidly transferred away from the region of self-focusing and, hence, defocusing occurs. In the AGVD regime, the anomalous GVD can still continue to transfer energy into the collapse region—even after plasma formation has occurred. This counter-balances the defocusing effect of the plasma. The consequence is that filaments in the AGVD regime do not spread so much and, as a result, longer filament lengths are expected before the beam eventually defocuses.

Furthermore, under AGVD conditions, the red components in the incident beam possess lower velocities than the blue components and, consequently, the frequencies that are generated during propagation are swept back into the main pulse. This results in a lower threshold for filamentation than in the normal GVD regime. Thus, the filamentation process is theoretically expected to give rise to a pulse that remains more or less invariant as it propagates: it is almost a spatio-temporal soliton. In practice, loss of a part of the optical energy does occur, with the conical emission (Fig. 6.1) being a major factor responsible for such loss [104]. Water has significant linear absorption in the AGVD regime and it is of biophotonic interest to explore whether, and how, white light generation from it may be affected by GVD considerations. SC generation in water has recently been explored over the wavelength range 800–1350 nm with incident power levels being kept low enough that only a single filament is formed [105, 106]; GVD values over this wavelength range cover the values from +24.8 fs^2mm^{-1} at 800 nm to -100 fs^2mm^{-1} at 1350 nm [107]. Typical spectra of white light obtained at these different wavelengths are shown in Fig. 6.10.

The linear absorption of 800 nm light by water is very weak (0.02 cm^{-1}) [108] and the SC spectrum is seen to extend over the range 350-1250 nm, corresponding to a span of about two octaves. Upon irradiation by 1250 nm light, the negative GVD regime is accessed and water now has larger linear absorption—of 0.9 cm^{-1}—and the resulting SC spectrum extends further into the infrared, to 1600 nm. The efficiency of SC generation over the wavelength range 400–1100 nm is ~8 % [105]. With even longer wavelength incident light (1300 nm), where the linear absorption is more significant (2.5 cm^{-1}), it requires a higher value of incident energy (5 μJ) to form a single filament; the extent of the SC spectrum now exceeds 2 octaves: it ranges from 350 to 1600 nm, and exhibits a noticeable dip at ~760 nm as well as a blue side continuum that peaks at 480 nm. The SC generation efficiency in this case is ~6 % [105]. At 1350 nm wavelength, water exhibits very significant absorption (5 cm^{-1}) and, as a result, incident energy values as high as 30 μJ are required to forma a single filament. The accompanying white light extends more than two octaves (350–1600 nm), with a prominent blue-shifted continuum which peaks at 450 nm and has a distinct dip at 710 nm. The blue-sided spectrum is narrower than the corresponding spectrum measured using 1300 nm incident light. The SC generation efficiency is reduced to 3 %.

Water offers the possibility of exploring a strong linear absorption regime at pump wavelengths where anomalous GVD values are readily accessed. From a biophotonics perspective, the results shown in Fig. 6.10 make clear that SC generation is a robust process despite the presence of substantial linear absorption which appears to neither preclude the formation of the blue side continuum nor the extent of the white light spectrum. "Engineering" of the spectral extent of the supercontinuum generated using water appears to be a distinct possibility simply by appropriate tuning of AGVD conditions—the wavelength of the incident laser pulses.

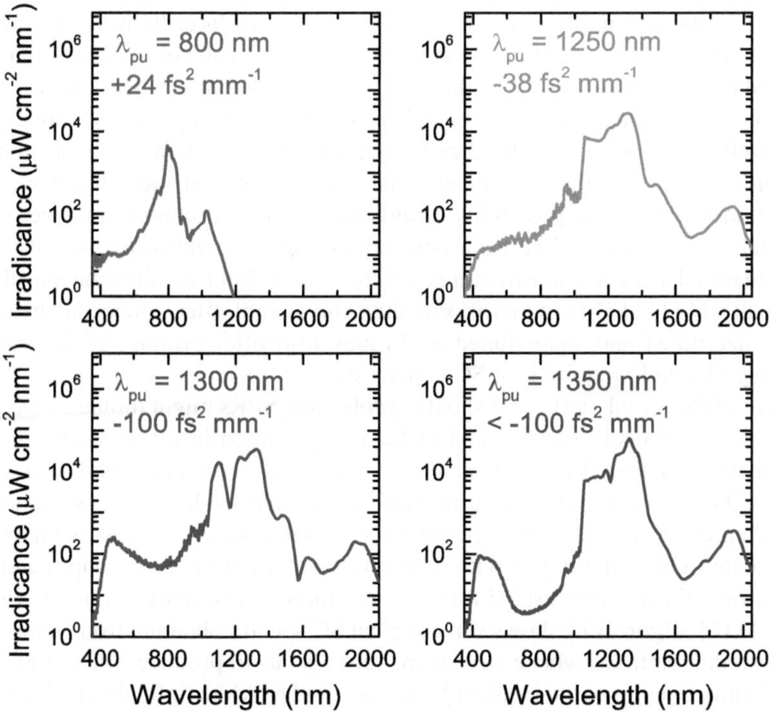

Fig. 6.10 Typical white light spectra upon irradiation of water by light of wavelengths: 800 nm (*top left panel*); 1200 nm (*top right panel*); 1300 nm (*lower left panel*); and 1350 nm (*lower right panel*). The incident energy was 3 μJ in the case of 800 nm and 1200 nm pulses, it was 5 μJ in the case of 1300 nm pulses and 30 μJ for 1350 pulses. Each of these energy values ensured a single-filament-regime

Interesting, and important, physics insights remain to be developed. It has been postulated that the blue side continuum peak observed in the vicinity of 480 nm may be amenable to rationalization as an isolated anti-Stokes wing resulting from interference of the SC light field as it undergoes anomalous group velocity dispersion. Such a process has been detected in fused silica [109]. An alternative rationalization involves the invocation of a three-wave mixing process in which linear dispersion plays a crucial role [105], with the spectral peak in the blue side being attributed to an axial component of conical emission [110].

6.3.3 Supercontinuum Generation in Water Doped with Nanoparticles

We have seen in Chaps. 3 and 4 how contemporary needs of imaging and microscopy have fueled efforts to probe nonlinear optical phenomena exerting wide ranging influence in photonic functionalities like the generation of ultrashort pulses and the possibility of controlling the frequency spectrum of laser light. However, many potential applications are often found to be limited by weak nonlinearities that, then, demand the use of laser pumping thresholds that are uncomfortably high from a biophotonics perspective. Enhancement of localized fields by optical techniques might effectively circumvent such hurdles by boosting the fast, non-resonant, nonlinear response of biologically relevant materials. In this context there has been recent interest in exploring the possibility of utilizing surface plasmon resonances (SPRs) in metal nanostructures, like nanorods and nanospheres embedded within biophotonic materials. Such nanostructures exhibit strong field localization on ultrafast timescales [111–113] potentially providing the sought-after enhancement mechanism. Size-tuned and shape-tuned SPRs may also offer promise of lowering the pumping threshold required for SC generation.

Most of the initial work carried out to probe how SPRs might influence SC generation was theoretical in nature and yielded only limited insights into mechanisms behind the anticipated enhancement of white light [114–117]. This was mainly because it proved difficult to experimentally realize the conditions that were assumed in the theoretical calculations. Experiments were subsequently initiated which probed SC generation in water doped with gold nanoparticles of different shapes and sizes, under irradiation by ultrafast (35 fs) laser excitation at very modest values of incident power [118]. Figure 6.11 shows some typical SC spectra obtained from water doped with Au nanoparticles whose sizes were varied so as to produce SPRs at 861, 617, and 554 nm. Spectra of the incident laser pulse and of white light obtained from pure (undoped) water are also shown. The SC spectra of doped water, covering the range 855–620 nm, show as much as 63 nm enhancement of spectral width compared to that in pure water. The broadening is more when the SPR overlaps the excitation laser's spectrum. It appears that judicious use of SPRs may prove useful for several applications involving imaging and microscopy—with only modest incident intensities (incident power levels less than $10P_{cr}$).

Note that introduction of dopants produced no significant change in the overall morphology of the filament produced in water (Fig. 6.11a) but the SC spectra revealed significant broadening that, in all cases, was asymmetrical around the incident laser wavelength of 800 nm. For $\lambda_R = 861$ nm, the spectral extent was large (173 nm) compared to the corresponding value (63 nm) obtained with undoped water. The peak of the SC spectrum was at 816 nm, somewhat blue-shifted. Conversely, when the dopant had a value of $\lambda_r = 617$ nm, the SC spectrum was much weaker although the spectral extent was not significantly narrower. A phenomenological model based on self-phase modulation accounts for these observations, taking into consideration contributions made by the plasma channel (filamentation) to the third-order susceptibility of water. The model also accounts for self-steepening of the falling edge of the ultrafast laser pulse—by introducing a time-dependent phase profile. The model, though simplified, successfully reproduces the observed spectral asymmetry around 800 nm [118].

Systematic measurements indicate that the width of the SC spectrum appears to be only weakly dependent on the position of the SPR (value of λ_r) while the enhancement in yield is both size as well as shape dependent. These attributes are of

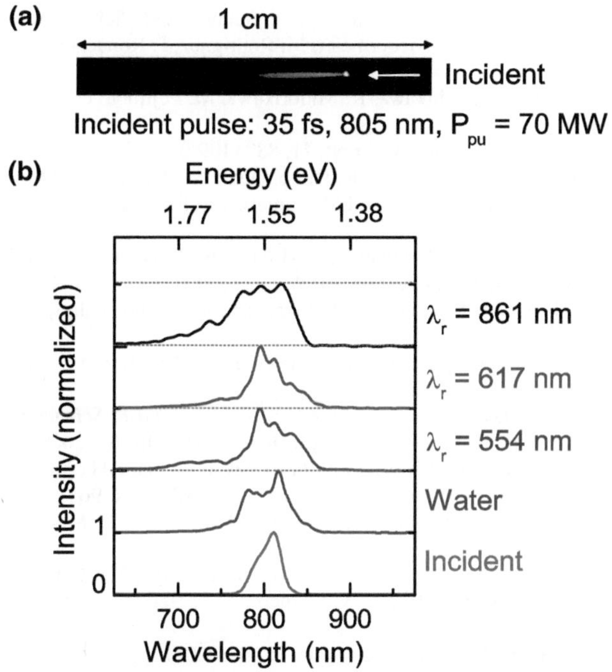

Fig. 6.11 **a** Image of a single filament formed in water irradiated by 35 fs long pulses of 800 nm light at an incident power of 60 MW. The water sample was doped with gold nanoparticles whose surface plasmon resonance, λ_r, was at 861 nm. **b** Typical white light spectra from water doped with Au nanoparticles with values of λ_r = 861, 617, and 554 nm. Spectra of the incident laser pulse and of white light obtained from pure (updoped) water are also shown

significance from the viewpoint of biophotonics as they afford yet another possibility of being able to "engineer" SC production by the relatively simple expedient of doping with nanoparticles of different sizes and shapes.

References

1. S.L. Chin, W. Liu, F. Théberge, Q. Luo, S.A. Hosseini, et al., in *Progress in Ultrafast Intense Laser Science III*, K. Yamanouchi, S.L. Chin, P. Agostini, G. Ferrante (Eds.) (Springer-Verlag, Berlin, 2008) p. 243
2. A. Couairon, A. Mysyrowicz, Phys. Rep. **441**, 47 (2007)
3. L. Berg, S. Skupin, R. Nuter, J. Kasparian, J.-P. Wolf, Rep. Prog. Phys. **70**, 1633 (2007)
4. M. Rodriguez, R. Sauerbrey, H. Wille, L. Wöste, T. Fujii et al., Opt. Lett. **27**, 772 (2002)
5. H. Pépin, D. Comptois, F. Vidal, C.Y. Chien, A. Despaarois et al., Phys. Plasmas **8**, 2532 (2001)
6. X.M. Zhao, J.-C. Diels, C.Y. Wang, J.M. Elizondo, IEEE J. Quantum Electron. **QE-31**, 599 (1995)
7. J.F. Gravel, Q. Luo, D. Boudreau, X.P. Tang, S.L. Chin, J. Anal. Chem. **76**, 4799 (2004)

8. J. Kasparin, M. Rodriguez, G. Méjean, J. Yu, E. Salmon et al., Science **301**, 61 (2003)
9. P. Rairoux, H. Schillinger, S. Niedermeier, M. Rodriguez, F. Ronneberger et al., Appl. Phys. B **71**, 573 (2000)
10. H. Kumagi, S.H. Cho, K. Ishikawa, K. Midorikawa, M. Fujimoto et al., J. Opt. Soc. Am. B **20**, 597 (2003)
11. E.N. Glenzer, E. Mazur, Appl. Phys. Lett. **71**, 882 (1997)
12. K.M. Davis, K. Miura, N. Sugimoto, K. Hirao, Opt. Lett. **21**, 1729 (1996)
13. L.C. Courrol, R.E. Samad, L. Gomez, I.M. Ranieri, S.L. Baldochi et al., Opt. Express **12**, 288 (2004)
14. K. Miura, J. Qiu, H. Inouye, T. Mitsuyu, K. Hirao, Appl. Phys. Lett. **71**, 3329 (1997)
15. H.B. Sun, Y. Xu, S. Juodkazis, K. Sun, M. Watanabe et al., Opt. Lett. **26**, 325 (2001)
16. M. Watanabe, H.-B. Sun, S. Juodkazis, T. Takahashi, S. Matsuo et al., Jpn. J. Appl. Phys. **37**, L1527 (1998)
17. E.N. Glenzer, M. Milosavljevic, L. Huang, R.J. Finlay, T.H. Her et al., Opt. Lett. **21**, 2023 (1996)
18. J.V. Moloney, M. Kolesik, in *Progress in Ultrafast Intense Laser Science II*, K Yamanouchi, S.L. Chin, P. Agostini, G. Ferrante (Eds.) (Springer-Verlag, Berlin, 2008) p. 253
19. K. Ishikawa, H. Kumagi, K. Midorikawa, Phys. Rev. E **66**, 056608 (2002)
20. R.Y. Chiao, E. Garmire, C.H. Townes, Phys. Rev. Lett. **13**, 479 (1964)
21. A.K. Dharmadhikari, F.A. Rajgara, N.C.S. Reddy, A.S. Sandhu, D. Mathur, Opt. Express **12**, 695 (2004)
22. Y.-H. Chen, S. Varma, T.M. Antonsen, H.M. Milchberg, Phys. Rev. Lett. **105**, 215005 (2010)
23. A. Couairon, J. Biegert, C.P. Hauri, W. Kornelis, F.W. Helbing et al., J. Mod. Opt. **53**, 75 (2006)
24. F. Wise, P. Di Trapani, Opt. Photonics News **13**, 28 (2002)
25. J.A. Dharmadhikari, R.A. Deshpande, A. Nath, K. Dota, D. Mathur, A.K. Dharmadhikari, Appl. Phys. B **117**, 471 (2014)
26. A. Braun, G. Korn, X. Liu, D. Du, J. Squier, G. Mourou, Opt. Lett. **20**, 73 (1995)
27. A.K. Dharmadhikari, J.A. Dharmadhikari, D. Mathur, Appl. Phys. B **94**, 259 (2009)
28. K. Dota, J.A. Dharmadhikari, D. Mathur, A.K. Dharmadhikari, Appl. Phys B. **107**, 703 (2012)
29. R.R. Alfano, *The Supercontinuum Laser Source* (Springer, New York, 1989)
30. R.L. Fork, C.V. Shank, C. Hirlimann, R. Yen, W.J. Tomlinson, Opt. Lett. **8**, 1 (1983)
31. P.B. Corkum, C. Rolland, T. Srinivasan-Rao, Phys. Rev. Lett. **57**, 2268 (1986)
32. J.T. Manassah, P.L. Baldeck, R.R. Alfano, Opt. Lett. **13**, 1090 (1988)
33. A. Penzkofer, A. Seilmeier, W. Kaiser, Opt. Commun. **14**, 363 (1975)
34. A.L. Gaeta, Phys. Rev. Lett. **84**, 3582 (2000)
35. N. Aközbek, M. Scalora, C.M. Bowden, S.L. Chin, Opt. Commun. **191**, 353 (2001)
36. X.J. Fang, T. Kobayashi, Appl. Phys. B **77**, 167 (2003)
37. D. Schumacher, Opt. Lett. **27**, 451 (2002)
38. N. Bloembergen, Opt. Commun. **8**, 285 (1973)
39. J.K. Ranka, R.W. Schirmer, A.L. Gaeta, Phys. Rev. Lett. **77**, 3783 (1996)
40. M. Kolesik, E.M. Wright, A. Becker, J.V. Moloney, Appl. Phys. B **85**, 531 (2006)
41. M. Kolesik, G. Katona, J.V. Moloney, E.M. Wright, Appl. Phys. B **77**, 185 (2003)
42. M. Kolesik, E.M. Wright, J.V. Moloney, Phys. Rev. Lett. **92**, 253901 (2004)
43. M. Kolesik, E.M. Wright, J.V. Moloney, Opt. Lett. **32**, 2816 (2007)
44. A.K. Dharmadhikari, F.A. Rajgara, D. Mathur, Appl. Phys. B **80**, 61 (2005)
45. A. Brodeur, S.L. Chin, J. Opt. Soc. Am. B **16**, 637 (1999)
46. A.K. Dharmadhikari, J.A. Dharmadhikari, F.A. Rajgara, D. Mathur, Opt. Express **16**, 7083 (2008)
47. A.K. Dharmadhikari, S. Edward, J.A. Dharmadhikari, D. Mathur, J. Phys. B **48**, 094012 (2015)
48. A.S. Sandhu, S. Banerjee, D. Goswami, Opt. Commun. **181**, 101 (2000)
49. A.M. Perelomov, V.S. Popov, M.V. Terentév, Sov. Phys. JETP **23**, 924 (1966)
50. S. Petit, A. Talebpour, A. Proulx, S.L. Chin, Opt. Commun. **175**, 323 (2000)

51. M. Kolesik, J.V. Moloney, E.M. Wright, Phys. Rev. E **64**, 046607 (2001)
52. P. Lambropoulos, Phys. Rev. Lett. **29**, 453 (1972)
53. W. Koechner, *Solid-State Laser Engineering*, 2nd edn. (Springer, Berlin, 1988)
54. H. Schroeder, J. Liu, S.L. Chin, Opt. Express **12**, 4768 (2004)
55. A.K. Dharmadhikari, F.A. Rajgara, D. Mathur, H. Schroeder, J. Liu, Opt. Express **13**, 8555 (2005)
56. A. Radzicka, R. Wolfenden, Science **267**, 90 (1995)
57. C. Bustamante, Z. Bryant, S.B. Smith, Nature **421**, 423 (2003)
58. X. Pan, P. Cloutier, D. Hunting, L. Sanche, Phys. Rev. Lett. **90**, 208102 (2003)
59. J.S. D'Souza, J.A. Dharmadhikari, A.K. Dharmadhikari, B.J. Rao, D. Mathur, Phys. Rev. Lett. **106**, 118101 (2011)
60. A.K. Dharmadhikari, H. Bharambe, J.A. Dharmadhikari, J.S. D'Souza, D. Mathur, Phys. Rev. Lett. **112**, 138105 (2014)
61. D.N. Nikogosyan, A.A. Oraevsky, V.I. Rupasov, Chem. Phys. **77**, 131 (1983)
62. R. Rudolph, K.-P. Francke, H. Miessner, Plasma Process Polym. **8**, 153 (2003)
63. D. Mathur, J.B. Hasted, Chem. Phys. Lett. **34**, 90 (1975)
64. H. Ramachandran, A.K. Dharmadhikari, K. Bambardekar, G. Ramanandan, J.A. Dharmadhikari et al., Nanotechnology **21**, 245102 (2010)
65. G. Ramanandan, A.K. Dharmadhikari, J.A. Dharmadhikari, H. Ramachandran, D. Mathur, Opt. Express **17**, 9614 (2009)
66. V.R. Bhardwaj, P.B. Corkum, D.M. Rayner, Phys. Rev. Lett. **93**, 043001 (2004)
67. J.B. Hasted, D. Mathur, in *Electron-Molecule Interactions and their Applications*, vol. 1, L.G. Christophorou, (Ed.) (Academic Press, New York, 1984), p. 403
68. D. Träutlein, M. Deibler, A. Leitenstorfer, E. Ferrando-May, Nucleic Acid Res. **38**, e14 (2010)
69. C. von Sonntag, *Free-Radical Induced DNA Damage and Its Repair* (Springer, Berlin, 2006)
70. M.A. Siddiqi, E. Bothe, Radiat. Res. **112**, 449 (1987)
71. R.E. Kirsch, M.B. Flick, C.N. Trumbore, Radiat. Res. **126**, 251 (1991)
72. M.L.T. Porro, M. Greenberg, J. Am. Chem. Soc. **135**, 16368 (2013)
73. K. Yuan, R.N. Dixon, X. Yang, Acc. Chem. Res. **44**, 369 (2011)
74. C. Santhosh, A.K. Dharmadhikari, K. Alti, J.A. Dharmadhikari, D. Mathur, J. Biomed. Opt. **12**, 020510 (2007)
75. U.M. Nater, R. La Marca, L. Florin, W. Langhans, M.M. Koller, U. Ehlert, Psychoneuroendocrinology **31**, 49 (2006)
76. A. van Stegeren, N. Rohleder, W. Everaerd, O.T. Wolf, Psychoneuroendocrinology **31**, 137 (2006)
77. S. Mishra, S.B. Noronha, G.K. Suraishkumar, Process Biochem. **40**, 1863 (2005)
78. M. Yamaguchi, M. Kanemaru, T. Kanemori, Y. Mizuno, Biosens. Bioelectron. **18**, 835 (2003)
79. F.L. Hofman, J. Nutr. **131**, 1621S (2001)
80. D.S.-Q. Koh, G.C.-H. Koh, Occup. Environ. Med. **64**, 202 (2007)
81. J.A. Bosch, H.S. Brand, T.J.M. Lightenberg, B. Bermond, J. Hoostraten et al., Psychosom. Med. **58**, 374 (1997)
82. D.M. Vasudevan, S. Sreekumari, *Textbook of Biochemistry for Dental Students* (Jaypee Bros. Medical Publishers, New Delhi, 2007)
83. G.M. Henningsen, J.J. Hurrell Jr., F. Baker, C. Douglas, B.A. MacKenzie et al., J. Work Environ. Health **18**, 133 (1992)
84. N. Rohleder, U.M. Nater, M. Jutta, U. Ehlert, C. Kirschbaum, Psychophysiology **43**, 645 (2006)
85. U.M. Nater, N. Rohleder, J. Gaab, S. Berger, A. Jud et al., Int. J. Psychophysiology **55**, 333 (2005)
86. G. Norhagen, P.E. Engström, L. Hammarström, P.Ö. Söder, C.I.E. Smith, J. Clin. Immunol. **9**, 279 (1989)
87. T. Otsuki, H. Sakaguchi, T. Hatayama, A. Takata, F. Hyodoh et al., Int. J. Immunopathol. Pharmacol. **17**, 45 (2004)
88. R.A. Martin, J.P. Dobbin, Int. J. Psychiatry Med. **18**, 93 (1988)

89. V. Kandidov, O.G. Kosareva, I.S. Golubtsov, W. Liu, A. Becker et al., Appl. Phys. B **77**, 149 (2003)
90. W.J. Jones, B.P. Stoicheff, Phys. Rev. Lett. **13**, 657 (1964)
91. C. Santhosh, A.K. Dharmadhikari, J.A. Dharmadhikari, K. Alti, D. Mathur, Appl. Phys. B **99**, 427 (2010)
92. C. Santhosh, A.K. Dharmadhikari, K. Alti, J.A. Dharmadhikari, D. Mathur, Proc. SPIE **6439**, 64390Q1 (2007)
93. R.A. Bank, E.H. Hettama, A.V. Amerongen, J.C. Pronk, Electrophoresis **12**, 74 (1991)
94. C.A. Burtis, E.R. Ashwood, D.E. Burns (eds.), *Tietz Textbook of Clinical Chemistry and Molecular Diagnostics*, 4th edn. (Saunders, New York, 2005)
95. N.H. Al-Rawi, F.A. Jaber, K.M. Atiyah, Third. World Med. **7**, 2 (2009)
96. R. Diab-Ladki, B. Pellat, R. Chachine, Oral Investig. **7**, 103 (2003)
97. M. Battino, M.S. Ferreiro, I. Gallardo, H.N. Newman, P. Bullon, J. Clin. Periodontol. **29**, 189 (2002)
98. M.L. Naudeau, R.J. Law, T.S. Luk, T.R. Nelson, S.M. Cameron, Opt. Express **14**, 6194 (2006)
99. A. Saliminia, S.L. Chin, R. Vallee, Opt. Express **13**, 5731 (2005)
100. M. Durand, K. Lim, V. Jukna, E. McKee, M. Baudelet et al., Phys. Rev. A **87**, 043820 (2013)
101. J. Darginavicius, D. Majus, V. Jukna, N. Garejev, G. Valiulis et al., Opt. Express **21**, 25210 (2013)
102. S.V. Chekalin, V.O. Kompanets, E.O. Smetanina, V.P. Kandidov, Quant. Electron. **43**, 326 (2013)
103. K.D. Moll, A.L. Gaeta, Opt. Lett. **29**, 995 (2004)
104. A. Houard, Y. Liu, A. Mysyrowicz, J. Phys. Conf. Ser. **497**, 012001 (2014)
105. P. Vasa, J.A. Dharmadhikari, A.K. Dharmadhikari, R. Sharma, M. Singh, D. Mathur, Phys. Rev. A **89**, 043834 (2014)
106. J.A. Dharmadhikari, D. Mathur, A.K. Dharmadhikari, in *Progress in Ultrafast Intense Laser Science XII*, K. Yamanouchi, L. Roso, R. Li, D. Mathur, D. Normand (Eds.) (Springer, Berlin, 2015) p. 105
107. A.G. van Engen, S.A. Diddams, T.S. Clement, Appl. Opt. **37**, 5679 (1998)
108. L. Bergé, Opt. Express **16**, 21529 (2008)
109. O. Smetanina, V.O. Kompanets, S.V. Chekalin, A.E. Dormidonov, V.P. Kandidov, Opt. Lett. **38**, 16 (2013)
110. M. Kolesik, E.M. Wright, J.V. Moloney, Opt. Express **13**, 10729 (2005)
111. M. Kauranen, A.V. Zayats, Nature Photon. **6**, 737 (2012)
112. L. Novotny, N. van Hulst, Nature Photon. **5**, 83 (2011)
113. S. Lal, S. Link, N.J. Halas, Nature Photon. **1**, 641 (2007)
114. C. Wang, Y. Fu, Z. Zhou, Y. Cheng, Z. Xu, Appl. Phys. Lett. **90**, 181119 (2007)
115. R. Driben, A. Husakou, J. Herrmann, Opt. Lett. **34**, 2132 (2009)
116. H. Baida, D. Mongin, D. Christofilos, G. Bachelier, A. Crut et al., Phys. Rev. Lett. **107**, 057402 (2011)
117. M. Pelton, M. Liu, S. Park, N. Scherer, P. Guyot-Sionnest, Phys. Rev. B **73**, 155419 (2006)
118. P. Vasa, M. Singh, R. Bernard, A.K. Dharmadhikari, J.A. Dharmadhikari, D. Mathur, Appl. Phys. Lett. **103**, 111109 (2013)

Chapter 7
Ultrafast Quantum Mechanical Processes in Plants

Abstract By conventional wisdom, quantum mechanical processes are not expected to manifest themselves in large systems like biological complexes at room temperature. However, ultrafast time-resolved spectroscopic investigations have yielded evidence to the contrary, suggesting that long-range quantum mechanical effects do occur in biological systems under physiological conditions. This chapter addresses some of the topics relevant to the functional role played by quantum processes in ultrafast biophotonic phenomena in plants along with their experimental investigations. We also briefly introduce the concept of quantum mechanical superposition and describe some important ultrafast spectroscopy techniques that may be used to explore coherent effects in biology.

7.1 Introduction

Are biological systems too complex to be easily described by the fundamental laws of physics and their associated mathematical models? The development, in the last decade or so, of several imaging and time-resolved techniques has led to the quest for physical and mathematical descriptions of microscopic biological systems and processes. Schrödinger's lecture series [1] and his famous book, "What is Life?", in which he correctly predicted some of the functional features of DNA, are amongst several well known examples of such early attempts [2–5]. Over the last decade, interdisciplinary approaches seeking consideration of biological systems in a wider, more quantitative, perspective have rapidly progressed, and an increasing number of branches of physics and mathematics are beginning to find applications in biology. These developments have raised questions like: Are quantum effects important in biology? Do they assist, perhaps enhance, a biological function? Since all atomic and molecular interactions rely on quantum mechanics, they certainly ought to be important from a microscopic viewpoint [1–3]. However, these effects have hitherto not been expected to manifest themselves on physiologically relevant length- and time-scales. Nevertheless, recent experimental evidence has begun to suggest that a variety of biological systems have succeeded—through evolutionary pressures—in harnessing some of the unique features of quantum mechanics to their advantage.

P. Vasa and D. Mathur, *Ultrafast Biophotonics*, Biological and Medical Physics, Biomedical Engineering, DOI 10.1007/978-3-319-39614-9_7

These features include coherent superposition of quantum states on physiologically important timescales and long-range tunneling [5, 6]. With hindsight, it is not surprising that an ordered arrangement of closely packed conjugated molecules will form the basis for the occurrence of these quantum effects.

One of the most important quantum effects—other than tunneling—is that a system can exist in a superposition of states. A classical system cannot. Therefore, we may consider a biological system that exploits coherent superpositions of states for some practical purpose to be the most direct evidence of quantum effects in biology. In this and the following chapter, we present a brief overview of phenomena occurring in plants and animals like (i) quantum effects in light-harvesting, (ii) charge transfer in photosynthesis, (iii) radical pair evolution in avian magnetoreception, and (iv) isomerization in early vision events. Some of these exploit quantum coherence, but not all. Particularly, in (i), (ii) and (iii), it is the persistence of coherent superposition that plays a central role [4, 5]. However, all of them involve displacement of energy and/or an electron wavepacket from one location to another through an energetically higher, thus a nonclassical, state instead of classical "downhill" hopping. Pictorially, it is like going *through* a wall instead of climbing it. Throughout this book, we have also explored biological functions in which quantum effects, like long-range tunneling of electrons through proteins and energy landscapes, play a vital role (Chaps. 1 and 9), but here quantum coherence is not involved. Since these effects generally occur on ultrafast timescale, several nonlinear spectroscopic techniques, mainly the different variants of two-dimensional (2D) spectroscopy, are being used, modified and developed to investigate coherent superpositions of quantum states and pathways [7, 8].

Perhaps the most intriguing and counterintuitive manifestation of quantum effects in biology is that of avian navigation in the earth's magnetic field. It appears to be the case that such navigation depends on exploiting the earth's magnetic field to tune radical pair production that aids both orientation and navigation. The possibility that complex biological systems perform what is essentially a kind of spin resonance experiment in order to navigate is certainly fascinating and is discussed in Chap. 8.

7.2 Quantum Coherence

In a simple two-level system, an atom or an exciton that makes a transition from an excited state to the ground state emits a photon with a characteristic frequency ω_0. This irreversible process is the most fundamental light-matter interaction and has been adequately described as a damped harmonic oscillator [9–11], as already discussed in Chap. 2. So far, we considered absorption and emission by only a single atom or a single molecule. Now, we look at what happens with an ensemble of molecules, as in an aggregate of biomolecules.

In general, even very weak interactions that occur before wavefunction overlap and covalent bonding take place can have dramatic effects on the optical response [12, 13]. The simplest system to study, yet one that is very instructive, is the physical dimer (two molecules bound together, usually by weak van der Waals forces). As long as the separation between monomers is larger than the size of a monomer, the point dipole approximation holds true and, consequently, a simple electrostatic interaction model can provide quantitative information about (i) the magnitude of the shift in resonance energy, (ii) the oscillator strength of dimer transitions, and (iii) the polarization of dimer transitions. However, quantum mechanics provides a more general description of the phenomenon, preserving predictions of the simpler electrostatic model. In the approximation of negligible ground-state interaction and only one of the monomers being in an excited state, the total Hamiltonian, $\widehat{H}$, of the coupled monomer system can be phenomenologically written in the form of a 2×2 matrix

$$\widehat{H} = \begin{pmatrix} H_{11} & V_{12} \\ V_{21} & H_{22} \end{pmatrix}. \tag{7.1}$$

Here, H_{ii} represents an individual monomer and $V_{12} = V_{21}^*$ represents the dipole coupling between two monomers [10, 11]. The coupled system is schematically depicted in Fig. 7.1a. Equation 7.1 provides a straightforward way of describing the linear optical response of the dimer. It incorporates some of the quantum mechanical effects within a semi-classical approach and can be easily extended to more advanced Bloch equations or the density matrix formalism. In presence of damping and disorder, $V_{12} \neq V_{21}^*$. Owing to its simplicity and wide scope, (7.1) has been

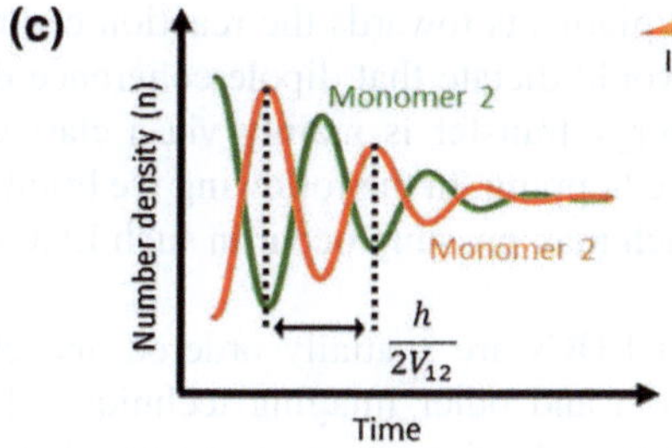

Fig. 7.1 **a** Schematic representation of dimer formation. **b** The dipole coupling V_{12} between the two monomers gives rise to two superposition states with energies E_+ and E_-. Excitation by a sufficiently short pulse can reveal the coupled system dynamics. **c** The dynamics are characterized by out-of-phase oscillations (period $\frac{h}{2V_{12}}$) in the monomer number density. In presence of damping, the oscillations are observed only during the coherence time of the system, which can be much shorter than the population relaxation time

successfully used to explain experimental results. Diagonalization of (7.1) leads to new states, $\psi_{\pm} = \frac{1}{\sqrt{2}}[\phi_1^*\phi_2 \pm \phi_1\phi_2^*]$ and respective energies, $E_{\pm} = \frac{H_{11}+H_{22}}{2} \pm V_{ij}$. The new states so obtained represent the coherent superposition of singly excited monomer states (Fig. 7.1b). Each of ϕ_i is a stationary solution associated with an individual monomer [5, 11, 14, 15]. The [+] and [−] states of the dipole model described above are the symmetric and antisymmetric wavefunction compositions. The dimer dynamics can now be described as the superposition of these symmetric and anti-symmetric wavefunctions (basis states).

The dipole interaction governing the formation of the coherent superposition—the entangled states—leads to a periodic transfer of energy between monomers via the emission and re-absorption of photons [11, 14, 15]. Thus, the nondiagonal coupling elements of V_{ij} represent the oscillatory terms that govern the quantum dynamics, at a characteristic angular frequency $\frac{2V_{ij}}{\hbar}$. As depicted in Fig. 7.1c, excitation by a sufficiently short laser pulse can reveal these coherent out-of-phase oscillations in the individual monomer's number density. In the absence of any damping or interaction with the surroundings (like solvent molecules), dephasing does not occur: the system continues to stay in the coherent superposition. In reality, it would decay at a rate determined by the interaction of the individual monomer with the surrounding and the nature of the coupling element, V_{ij}. Thus quantum coherence, when associated with a dissipative mechanism, leads to energy transfer between individual systems as well as to the surrounding [11, 14, 15].

Our aim here is to answer the question: Is there any biological process that relies on quantum mechanics to perform a task that either cannot be carried out classically, or is carried out more efficiently than the best classical equivalent? We qualitatively describe how coherent energy transfer involving entangled states is much more efficient than the classical incoherent energy hopping.

A typical example of an energy transport mechanism in biophotonics is that of light-harvesting complexes (LHCs) in plants and photosynthetic bacteria [4, 5, 12, 16–18], as shown in Fig. 7.2 These complexes comprise photosynthetic pigments and proteins surrounding a photosynthetic reaction centre. They funnel the photon energy absorbed by the antenna pigments towards the reaction centre via a bridge system. Conventional wisdom would dictate that dipole coherence decays quickly in such systems and, hence, energy transfer is mainly via a classical "downhill" mechanism mediated by inter-site hopping. In the following we briefly discuss three energy transfer mechanisms which may possibly occur in such LHCs [4, 5, 12, 16, 19–21].

The constituent molecules of LHCs are spatially ordered and closely packed, as confirmed by X-ray diffraction and other imaging techniques [4, 5, 21, 22]. Their interaction might, thus, lead to the formation of entangled states supporting resonant energy transfer. As shown in Fig. 7.2a, the energy from the antenna system, An, is transferred to Br_i, which is one of eight states in the bridge system, Br_n, via Förster type energy transfers [4, 5, 22, 23], where the energy transfer between states occurs incoherently. Due to the dipole interaction, Br_i is not isolated but is coherently coupled to the other sites in the bridge system. The photon energy from

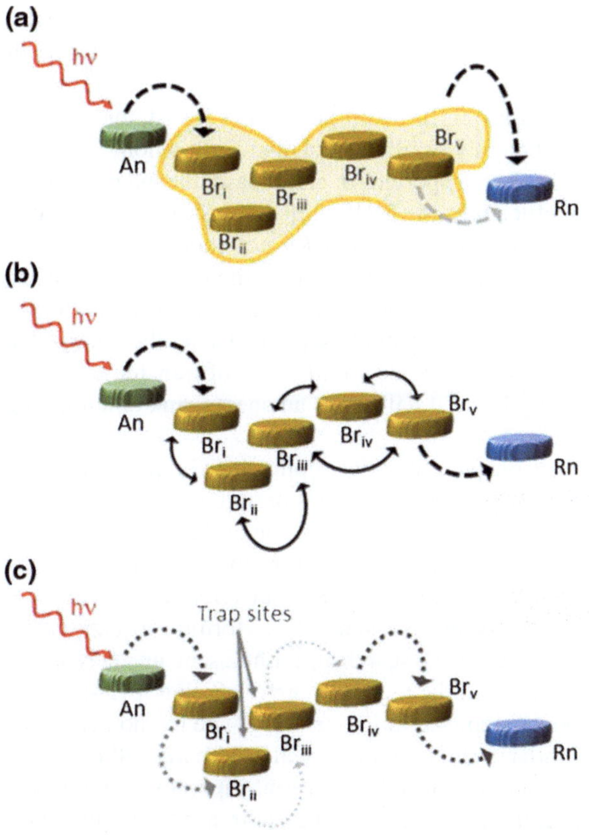

Fig. 7.2 Energy transfer in a typical photosynthetic light-harvesting complex (LHC). **a** Energy transfer from an antenna site, An, consisting of a photosynthetic pigment to a final photosynthetic reaction centre, Rn, via the coherent superposition of bridge sites, Br_n (only five of eight are shown). The x-coordinate is the reaction coordinate; the y-axis denotes energy. The shaded area defines the extent of the coherent superposition of the bridge states. The *dashed*, *solid* and *dotted arrows* represent energy transport via Förster type, coherent and inter-site hopping, respectively. Interestingly, the energy is efficiently funneled from An to Rn even though there are lower-energy intermediate bridge sites, Br_{ii} and Br_{iii}. **b** Similar to **a** but the energy oscillates back and forth among the bridge system sites before being transferred to Rn. **c** Classical "downhill" energy transfer between An and Rn by inter-site hopping via the bridge system. The lower energy sites Br_{ii} and Br_{iii} act as energy sinks, hindering the flow of energy (*grey dotted arrows*), thus yielding a very inefficient transport mechanism

An is thus transferred to a collective state, that is, a coherent superposition of local wavefunctions of the bridge system. As there is strong coherence between various sites (Fig. 7.2a), even if only one of these sites, for instance, Br_v, is relatively strongly coupled to the reaction centre, Rn, it will result in energy transfer from An to Rn.

On what timescale does such energy transfer occur? This depends on the strengths of the $An - Br_i$ and $Br - Br$ couplings. If the coupling is weak, the initial energy transfer is slow and only the collective state, as in Fig. 7.2a, is populated at the end of the $An \rightarrow Br_i$ transfer. If the initial transfer is fast, and the coupling is strong, the energy will coherently oscillate back and forth among various sites (Fig. 7.2b) before it is completely transported. In either case, the energy reaches Br_υ and then, finally, Rn very efficiently [4, 5, 19–23]. Interestingly, in this picture the energy transfer may occur in spite of An and Rn not being strongly coupled to the same site in the bridge system. Also, each bridge site could be coupled to several other sites but only those which are nearly resonant will receive a large fraction of the energy, even though they are spatially separated and other energetically lower states are available. In the classical "downhill" type of transfer, even if one of the sites, Br_{ii}, is energetically lower it will act as an energy sink or a trapping site, hindering energy transfer (Fig. 7.2c).

A more microscopic picture of the implications of quantum coherence for energy transfer suggests that superposition states permit excitation energy to flow back and forth (Fig. 7.2b). Thus once transferred to one of the Br_i sites, the system can simultaneously sample energy exchange rates from all constituent states and can efficiently direct energy transfer to the most effective site (Br_υ) without any loss [5, 19]. When viewed in this way, the system is essentially performing a quantum computation similar to that proposed in Grover's search algorithm: sensing many states simultaneously and selecting the one with the highest transfer efficiency [24]. In inter-site hopping, the excitation moves stepwise from one bridge site to another, dissipating energy at each step—in similar fashion to a classical search algorithm, where only one state can be occupied at a time. Thus, the coherent superposition of states funnels energy much more efficiently than would incoherent transport through the bridge sites.

For many years, it was believed that in complex biological systems coherence would not survive for more than a few femtoseconds. As a consequence, coherent processes were disregarded [4, 5, 19–23]. However, ultrafast spectroscopy is beginning to reveal the relevance of coherent processes in biological systems.

7.3 Ultrafast Spectroscopy

Though entanglement is beginning to be accepted as a natural feature of coherent evolution, it continues to be viewed as a fragile and exotic property of isolated systems in pure states. However, the occurrence of entanglement is more widespread than previously anticipated: it exists in biological systems under physiological surroundings [25]. Studies of coherent effects in ultrafast biophotonics are attracting attention, with the primary questions being: Can we detect entanglement? Can we quantify coherent effects? The important parameter in this context is the extent and nature of damping as it rules the entanglement and dictates the lifetime of the coherent state. Typical lifetimes of coherent states in biological systems are $\sim$100 fs [4, 5, 19–23]. Hence, time-resolved studies of such coherent effects need recourse to ultrashort pulses: the shorter the pulse, the better would be the resolution. We describe two such endeavours (for detailed discussions, see [7, 8, 26–28]).

The simplest time-resolved technique to probe coherent dynamics is pump-probe spectroscopy [29–31]. Here, the pulse train from an ultrafast laser is divided into two: the biosample is excited by one pulse train (the pump, which is generally more intense) and the changes it induces in the sample are probed by the second pulse train (the probe, generally weaker), which is suitably delayed with respect to the pump. A property of the probe (typically, transmission or reflectivity) is monitored to investigate the changes produced by the pump in the biosample. The resulting differential transmission (or reflectivity) maps, $\Delta T/T(\tau_{pp}, \omega_{pr}) = (T_{on} - T)/T$, can yield detailed information about the dynamics of the biosystem being studied. Here, ω_{pr} is the probe frequency and T_{on} (T) is the transmitted or reflected probe pulse spectrum in the presence (absence) of the pump pulse, recorded as a function of the delay, τ_{pp}, between pump and probe pulses. Pump-induced nonlinear optical processes—like saturation of absorption, excited state absorption, Stark shift and stimulated Raman scattering—are some of the mechanisms that govern the transient differential signal. Depending upon the characteristics of the pump/probe pulses and the spectral quantity being monitored, several variants of this technique have been developed [26–32]. To investigate coherent dynamics, the pump and probe pulse spectra and polarization have to be identical and the pulse duration has to be shorter than the coherence lifetime of the system. For an isolated system, like an isolated LHC, or one with few optical transitions, pump-probe spectroscopy can provide information about the coherence as well as the excited state population dynamics. In Chap. 4, we have discussed investigation of coherent energy transfer dynamics in a single LHC using pump-probe spectroscopy with shaped pulses. However, for biosystems, it might not be able to distinguish if the two or more observed resonances are coupled or not. Also, in presence of multiple damping mechanisms, it becomes difficult to identify individual contributions.

A signature of the correlated dynamics in a composite biomolecule is a splitting of the characteristic absorption spectrum of the composite. Such changes are very challenging to observe because of two primary difficulties: (i) due to the ultrafast nature of the photochemical reaction, the measured absorption spectrum changes transiently on ultrafast timescales, and (ii) due to the presence of other spectral broadening mechanisms, like inhomogeneous broadening, the splitting is masked. Ultrafast two-dimensional spectroscopy [7, 8, 19, 20], schematically shown in Fig. 7.3a, is an emerging technique that seeks to address both these difficulties. It allows tracking of spectral changes in real-time; it also offers a possibility to separate contributions from incoherent (inhomogeneous broadening in an ensemble) processes from quantum coherence and permits measurement and imaging of the non-diagonal coupling elements of the Hamiltonian of a multilevel system. Essentially, it is akin to real-time imaging of the density matrix of a biosystem.

Unlike usual pump-probe techniques, 2D spectroscopy utilizes three excitation pulses to induce nonlinear polarization in a sample [33, 34]. An example of a 2D pulse sequence is shown in Fig. 7.3a. Two pump pulses are used to excite the biosample. The time delay between these pulses (τ) is generally within the coherence time of the

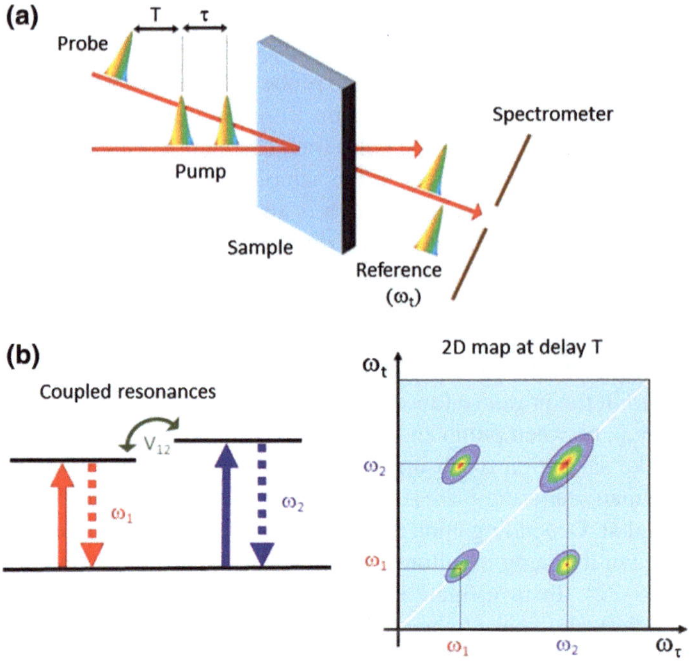

Fig. 7.3 **a** A set-up for 2D spectroscopy. A pump-pulse pair (separated by τ) excites the biosample and resulting changes are monitored by a delayed (T) probe pulse. The transmitted probe pulse is heterodyned with a reference pulse to get ω_t. By varying τ and T, information about coherent dynamics can be obtained. **b** A typical 2D map of a system comprising two coupled resonances. The diagonal peaks correspond to the individual resonances, whereas the cross-peaks are due to the coupling V_{12}

system. At some time T after the pump pulses, a third pulse probes the sample, and the emitted field provides information about how the system has evolved. The emitted field is often detected by heterodyning with a reference pulse in order to retrieve frequency and phase information. The heterodyned signal is usually collected in the frequency domain by a spectrometer, yielding ω_t. In all these measurements phase stability between the pump pulses has to be preserved. Several approaches have been attempted to simplify the generation of phase stabilized pulse sequences including the use of birefringent wedges [35–39]. Upon Fourier transforming the recorded spectra, the delay axis (τ) is correlated to excitation and emission energies (a 2D map) of the constituent sub-systems at a given pump-probe delay T. Other variants involving three excitation pulses but no heterodyning include 4-wave mixing and photon-echo spectroscopy [35, 36, 40, 41]. Implementation of pulse-shaping techniques have also made it possible to adapt 2D spectroscopy to other multidimensional methods [42], including extension into the UV regime [43, 44]. With the development of high-intensity coherent X-ray sources (Chap. 11), multidimensional X-ray spectroscopy [45, 46], is expected to provide promising new perspectives on electronic structure, energy-transfer and charge-transfer processes.

Figure 7.3b depicts two coupled resonances, and the resulting map exhibiting diagonal peaks corresponding to excitation of and emission from individual resonances. Here, coupling ensures that excitation of one of the resonances also results in emission by the other, giving rise to cross-peaks (Fig. 7.3b); these are a characteristic signature of the coupling. Apart from real-time imaging of coupling, this technique can also help estimate the extent of inhomogeneous broadening in an ensemble. Independent control over the polarization and spectrum of the three excitation pulses enables studies of more complex correlations and provides information about the orientation of transition dipole moments. Among many other biologically relevant proteins and composite molecules, LHCs have been successfully investigated using 2D spectroscopy [19, 47].

7.4 Ultrafast Light-Harvesting and Energy Transport in Photosynthesis

During photosynthesis the light energy, mainly from our sun, is absorbed by light-harvesting antennas and is then transported to a reaction centre where charge separation creates stable and storable chemical energy [4, 5, 19–23]. Though the composition and architecture of plants and organisms vary, the overall function of the photosynthetic system—to create a charge separated state [4, 5, 16–18] with $\sim$100 % quantum efficiency—is surprisingly similar! Light-induced processes of many photosynthetic organisms and pigments have been studied and two limits of excitation energy transport have been identified: (i) incoherent Förster type or hopping, and (ii) coherent phonon-induced relaxation between exciton levels. The important factor in deciding which of the two processes will dominate is the ratio of dipole coupling, V_{ij}, to the disorder, Δ. The disorder may be either due to damping or competing processes like electron-phonon interactions.

$\frac{V_{ij}}{\Delta} \ll 1$ defines the weak coupling regime and Förster transfer dominates. $\frac{V_{ij}}{\Delta} \geq 1$ corresponds to the strong coupling regime and here, a coherent superposition occurs and the wavefunctions of the system's energy eigenstates, usually called Frenkel excitons [16], are primarily delocalized across the interacting pigment molecules. While both limits can be described by quantum mechanics, it is the real experimental situation that decides the choice of basis (molecular or excitonic) states. The ratio $\frac{V_{ij}}{\Delta}$ also determines the extent of exciton delocalization and system dynamics. Generally, both energy transport mechanisms coexist and both have to be accounted for in developing microscopic understanding of the transport mechanism [48–51].

7.4.1 The Structure of the LHC: An Efficient Quantum Machine

The precise biological structure and pigment constituent of LHCs vary between organisms. For example, purple bacteria use highly symmetric ring-like structures

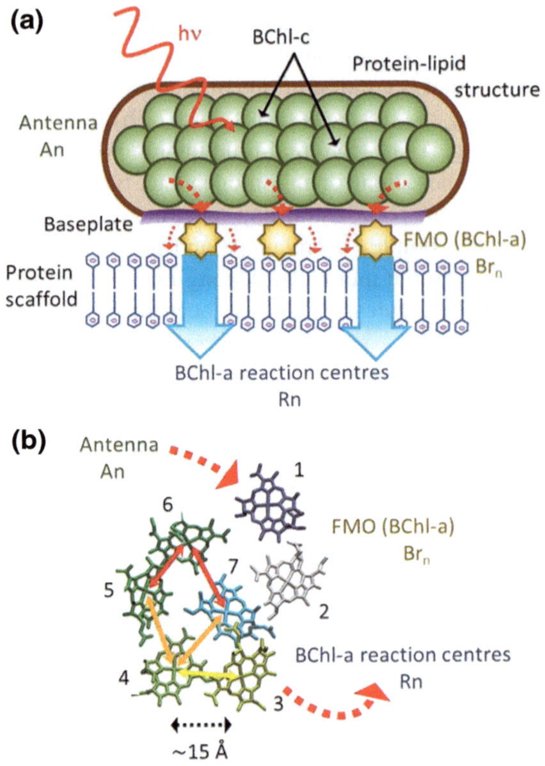

Fig. 7.4 Highly efficient photosynthetic quantum machinery of green sulphur bacteria. **a** The schematic illustrates absorption of light by the chlorosome antenna (An) and transport of the resulting excitation to the reaction centre through the FMO protein (Br_n). The exceptionally large chlorosome antenna (*green circles*) is composed of roughly ~200,000 bacteriochlorophyll-c (BChl-c) molecules encased in a lipid-protein structure to facilitate capture of as many photons as possible. The photo-excitation in this antenna is transferred (*red dashed arrows*) to the reaction centre (Rn) via one of several FMO complexes connected to the baseplate. **b** The BChl-a arrangement of one of the FMO monomers obtained by X-ray diffraction. It comprises eight (although only seven are shown here) closely packed BChl-a molecules encased in a protein scaffold. Some pairs of chromophores are moderately strongly coupled to form dimers (*coloured arrows*). The excitation arrives from the chlorosome at one of the sites, typically thought to be the site denoted as 1 (represented as Br_i in Fig. 7.2). This excitation is then coherently transported to coupled BChl-a molecules. Once it arrives at site 3 connected to Rn (represented as Br_υ in Fig. 7.2), it irreversibly enters the reaction centre to start a charge-separation process

for light harvesting, whereas green plants and cyanobacteria have LHCs with randomly arranged chlorophyll pigment molecules [4, 5, 16–18]. Such diversity reflects the necessity for organisms to adapt in response to different natural habitats and physiological conditions. One of the simplest and most-studied examples is the LHC of green-sulphur bacteria [4, 5, 19, 21, 52], schematically shown in Fig. 7.4a and briefly discussed in Sect. 7.2. This LHC has a very large chlorosome antenna protein

(An), consisting of ~200,000 bacteriochlorophyll-c (BChl-c) molecules encased in a protein-lipid structure, for capturing sunlight and thriving even in low-light conditions. The energy collected by these chlorosomes is transferred to the reaction centre through a specialized structure called the Fenna-Matthews-Olson (FMO) complex (bridge system, Br_n). The reaction centre (Rn) protein is nature's version of a solar cell; it carries out charge separation and transfer. What is remarkable is the observed efficiency of this LHC. Almost *every* absorbed photon is successfully transported to Rn, even though the electronic excitations at An and Br_n are very short-lived (~0.5 ns). If not transferred much faster than ~0.5 ns, the energy is lost via luminescence [4, 22].

In 2007, Engel and co-workers obtained evidence for quantum coherent energy transfer in the FMO complex using ultrafast 2D-IR spectroscopy and since then, the FMO protein has become one of the mainstays of quantum biology [19–23, 34]. Owing to its relatively small size and solubility in water, it has also been well characterized using X-ray diffraction [4, 5, 21]. The FMO complex normally exists in a trimer of three independent complexes, each of which consist of eight BChl-a molecules bound to a protein scaffold (Fig. 7.4b). The latter is the primary source of decoherence and noise but may also assist in protecting coherent excitations (Chap. 4) in the complex; it plays a role in promoting high transport efficiency. The complex is connected to the chlorosome antenna through what is called a baseplate. Excitations enter the FMO complex from this baseplate, transferring one of the closely packed BChl-a molecules (represented as Br_i in Fig. 7.2) into its first singlet excited state. The structure of the FMO Hamiltonian indicates that some pairs of chromophores are reasonably strongly coupled (owing to their close proximity with inter-molecular separation of only ~15 Å and favourable dipole orientations) and, hence, they effectively form dimers [4, 21, 53, 54]. The chromophore wavefunctions, or Frenkel excitons, are primarily delocalized across these dimers [16, 55, 56]. One of the BChl-a molecules (represented as Br_v in Fig. 7.2) is connected to Rn. Upon reaching there, the energy is irreversibly transported to the reaction centre for further processing [4, 21].

7.4.2 Quantum Energy Transport

The FMO complex is an unusual light-harvesting component because it acts primarily as a wire, to transport excitations between the light-harvesting antenna of green sulphur bacteria and their reaction centres [4, 21, 52]. Most other light-harvesting structures, like the antenna complexes associated with the Photosystem II (PSII) reaction centres in green plants and algae that drive the water-splitting reaction, play dual roles in light capture and excitation transport [5, 22]. Direct evidence of quantum coherence over appreciable length scales and timescales in the FMO complex was obtained in 2D spectroscopy of coherent electronic oscillations across multiple pigments within the FMO complex at low temperature (77 K). Surprisingly, the quantum beating was observed to last as long as 660 fs [19], in stark contrast to the general assumption that quantum coherences would be rapidly destroyed in complex systems. Several studies have now confirmed that coherence is non-negligible even at room temperature, for up to 300 fs [34, 57]. Other examples of long-lasting coherence include the main peripheral as well as core light-harvesting antennae in purple bacteria [known as light-harvesting complex 2 (LH2) discussed in Chap. 4, and light-harvesting complex 1 (LH1), respectively], where measurements of coherent and/or super-radiant emission [56, 58] suggest that entangled states persist for the lifetime of the excited state or for the time taken to transfer the excitation to Rn.

The notion of preserving entanglement in complex systems is a challenge for contemporary science. In this context, it is astonishing that this fragile quantum property survives for $\sim$300 fs in a warm, wet and complicated biological system [34, 57]. However, the key point is how strongly the quantum systems are "wired" together [4, 21, 52]. For FMO it is instructive to compare thermal energy at room temperature (300 K$\sim$26 meV$\sim$160 fs) with the coupling strength between the excitation in the FMO complex and the protein environment ($\sim$12 meV$\sim$350 fs) and the electronic coupling strength between BChl-a molecules ($\sim$12 meV). These values are comparable [59, 60]. Therefore quantum effects would, indeed, be important at room temperature [34, 57]. Deduction of the energies and coupling strengths in FMO requires a combination of quantum chemistry and ultrafast spectroscopy. Fortunately, FMO is well-studied, and theoretically estimated coupling strengths agree quantitatively with measured values [54, 59, 60]. A density matrix study of entanglement in the FMO complex and its temperature dependence [21] has revealed that entanglement in LHCs depends on the extent of delocalization of electronic excitation. Under physiological conditions it is predicted that entanglement, mainly between chromophore dimers, persists essentially for the lifetime of the excited state [21]. It is also suggested that finite long-range coherent delocalization of photoexcitation may exist across several molecular sites [19, 21].

It can be appreciated that a comprehensive theoretical model would need to include wave-like energy motion (owing to long-lived coherence terms), electron-phonon interactions, vibronic transitions, and other non-Markovian processes alongside the usual population transfer [4, 21, 22, 61]. The role of lattice vibrations within photosynthetic proteins may not be limited to simply passively encasing the coupled chromophores; they may actually modulate such couplings by altering the local dielectric environment such that exciton energies shift to promote coherent transfer [19, 21]. Some of the theoretical approaches mentioned here are based on the quantum master equation describing the time evolution of reduced density matrices. The ensemble (statistical) average inherent in such an approach may wash out details of the quantum behaviour [62, 63]. Other theoretical investigations show that quantum-coherent motion is robust in the individual realizations of the environment-induced fluctuations, contrary to the intuition obtained from the reduced density matrices [64, 65]. These results imply that experimental observations of delocalized states indicate the existence of wave-like energy flow in individual complexes (Chap. 4) and not in an ensemble. Based on these preliminary results, it would be very rewarding to investigate quantum phenomena at the individual protein level for a variety of biological processes. Experimentally, ultrafast tomography on LHCs is not yet possible, but techniques like ultrafast electron microscopy (Chap. 3) might enable

3D-imaging of the LHC in real-time and enable extraction of individual coupling elements. It has recently been demonstrated that polarization-dependent ultrafast 2D spectroscopy [47, 66] can determine the orientation dependent excitonic dipole moments and, consequently, dynamical estimation of density matrix elements is within reach!

If quantum coherent dynamics are indeed present at room temperature in the FMO complex (and other parts of a light-harvesting apparatus), what purpose do they serve? A higher transport efficiency is a plausible answer [67]. Many models have sought to address if, how, and why nature relies on quantum coherence to "move" electronic excitation through the FMO complex [61]. It is not immediately obvious that light-gathering efficiency is necessarily improved by quantum coherence, but it could be the resistance to "trap states", the robustness to survive under different conditions, or improved regulation of light harvesting which, via evolutionary pressures, would have helped organisms thrive in a low-light conditions [34, 61, 68, 69].

Though existence of quantum coherence is now widely accepted, alternative explanations exist that do not fully rely on entanglement [70–73]. These focus on different types of protein environments surrounding the FMO complex, or treating the protein environment as a markovian and uncorrelated thermal bath in which each site in the FMO complex is subjected to individual random environmental fluctuations [61, 74, 75]. A combination of an excitonic state and a rapid incoherent transfer to the reaction centre has also been suggested [76]. However, because of the strong coupling ($\sim$12 meV) between electronic excitations and protein environments [59, 60], the consensus is that models that treat each site individually are inadequate [77, 78]. Despite the positive results and predictions we have discussed, recent analysis has indicated that benefits in efficiency provided by quantum models may be only a few per cent [79]; nonetheless, even a few per cent improvement in overall efficiency may be vital in an organism's attempt to survive in low-light conditions [4, 21].

Only very few in vivo observations of coherence have been performed [80, 81]. Experimentally, excitations are induced using laser pulses with a high degree of coherence. In vivo, excitations are generated by incoherent sunlight or through energy transfer from another antenna complex. Does quantum coherence still play a significant role in such a situation? Most experimental and theoretical results that we have discussed have omitted the recently discovered eighth BChl-a molecule. There are contradictory reports about its constructive role in coherent transport [82–84]. In higher plants the photosynthetic apparatus is more complex and less ordered, suggesting that the energy landscape must be very rugged and adaptive [34]. Thus, it could be that the quantum coherence seen in FMO complex plays an even more important role in complex systems. Observation and validation of quantum transport is one of the big open problems in ultrafast biophotonics. A broader understanding of the role of coherence in a larger range of photosynthetic complexes and organisms (for example, LH-1, LH-2, LHCII, and carotenoids) is urgently needed [85–87].

7.5 Charge Separation and Transfer in Photosynthesis

Photosynthesis is a complex and remarkable process which is able to adapt to a fluctuating environment and is also able to self-regulate [16, 17, 88]. Understanding microscopic mechanisms involved in it is a prime example of research that requires the combined effort of several disciplines, including quantum mechanics, biophysics, biochemistry, molecular and structural biology, as well as physiology and ecology [3, 5]. The timescales involved in photosynthetic reactions range from femtoseconds to days, which again highlights the complexity of this process [5, 16, 17, 89, 90].

As already noted, an extensive body of evidence has lead to the conclusion that photosynthetic organisms have developed extraordinarily efficient mechanisms for harvesting solar energy; elaborate LHCs collect solar energy and transfer it to photochemical reaction centres with almost 100 % quantum efficiency within ~1 ps [5, 16, 17]. Once inside the reaction centres, the photon energy induces charge separation that, in turn, drives slower biochemical reactions, ultimately producing chemical energy. At the heart of a reaction centre is an electrochemical process involving three coupled electronic states corresponding to the excited bacteriochlorophyll or chlorophyll "special pair" (donor), the reduced bacteriopheophytin or plastocyanin (acceptor), and the reduced accessory chlorophyll or bridging chlorophyll [89–92]. The "special pair" (P870 in purple bacteria, P700 in photosystem-I or PSI and P680 in PSII of plants) molecules are very different from antenna chlorophylls. When they receive energy from the antenna pigments, they generate a redox-active chemical species with characteristic spectroscopic properties. Excited P680* (or P870* in purple bacteria) undergo oxidation ($P680^+$ or $P870^+$) to donate an electron to other bridging components and then, eventually, to the plastocyanin and to PSI (bacteriopheophytin in purple bacteria). The electron transferred by plastocyanin in plants is finally accepted by the oxidized state, $P700^+$ in PSI. The charge separation in PSI and PSII, together with the electron transfer through the bridging complexes, leads to the generation of an electrochemical potential gradient (the sum of the chemical potential due to differences in concentration and the electrical potential due to the variation in charge-concentration), which powers ATP synthesis by the protein complex, F-ATPase [89, 93, 94]. The "special" property of this arrangement lies in its ability to induce electron transfer via a series of proteins within ~200 ps to drive ATP synthesis. Efficient charge separation occurs only if the photon is absorbed close to the reaction centre. However, reaction centre pigments like P680, P700 and P870 have relatively lower absorption cross-section, requiring the presence of LHCs for efficient light absorption.

The remarkable efficiency, adaptivity and directionality of electron transfer in the reaction centre are all essential for photosynthesis to work. Dense and precise packing of the participating proteins is essential, as the the resonant energy transfer efficiency strongly depends on the distance between the chromophores [89, 90]. In recent years, several attempts have been made to mimic this highly sought-after chemical catalyst for light-driven charge separation in artificial photovoltaics (Chap. 10). The architecture of the light-harvesting and reaction centre complexes contains important secrets about how to efficiently capture, transfer, convert, and store solar energy [89]. Ultrafast spectroscopic studies have focussed on understanding the design principles of photosynthetic systems but they continue to present challenges: the choice of pigments, their particular protein environments, relative distances, and orientations produce broad inhomogeneous transitions that reflect varying degrees of electronic coupling, disorder, and dynamical fluctuations of the protein matrices. The ultrafast nature of the energy-transfer and charge separation processes further complicates an already demanding spectroscopic problem. As already noted, 2D visible and/or IR spectroscopy has emerged as a valuable tool for probing the structure-function relationship in natural photosynthetic complexes [19–21, 23, 34, 42, 95].

7.5.1 PSII Reaction Centre Complex

Recent efforts aimed at structural determination of photosynthetic complexes have provided a comprehensive description of the architecture of energy transduction in oxygenic photosynthesis that has been sought for several decades. Structural and spectroscopic insights into these complexes have yielded an accurate framework for the exploration of processes as well as of the evolutionary pressures that shaped the photosynthetic apparatus of various organisms [89, 90]. All four protein complexes that are necessary for the light-driven reactions of photosynthesis (light-harvesting and electron transport) in green plants reside in chloroplasts, organelles that are bound by a double membrane structure (Fig. 7.5). The protein cofactors are con-

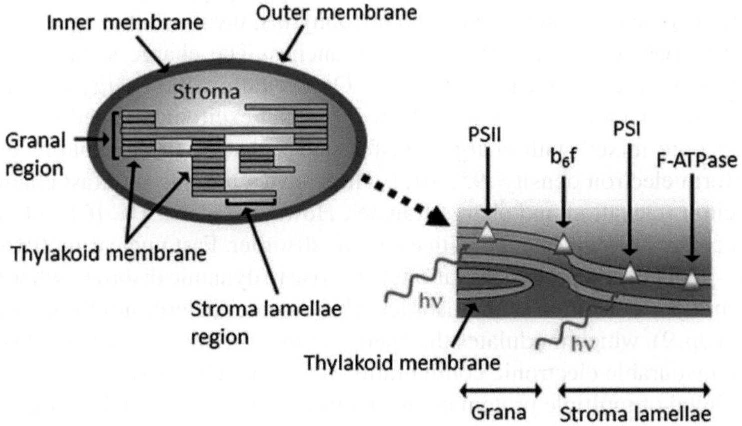

Fig. 7.5 A schematic depiction of the distribution of the four photosynthetic protein complexes, PSI, PSII, cytochrome b_6f and ATP synthase (F-ATPase), in the chloroplast of green plants and algae. Chloroplasts are organelles (~5 μm) bound by a double membrane structure. The third membrane, the thylakoid contains four unevenly distributed protein complexes. The regions consisting of stacks of thylokoids are called grana, which are connected by a non-stacked membrane called the stroma lamellae. The fluid surrounding the thylakoids is the stroma. The principal water splitting reaction involving charge separation occurs in PSII

tained in a third chloroplast membrane known as thylakoid. As depicted in Fig. 7.5, the cofactors have a specific location and are not evenly distributed throughout thylakoids. The two photosystems PSI and PSII containing numerous pigments for light harvesting and reaction centres have evolved to operate with a high quantum yield that is unmatched by any artificial system. Though found in the same organelle, PSI and PSII have different structure, function and quantum yields. Understanding the link between structure and function in this unique system still remains a challenge in ultrafast biophotonics. Though the steady state structures of PSI and PSII have already been extensively studied using electron microscopy [96, 97], knowledge of the dynamical structural changes of pigments and their distributions, as obtained by time-resolved X-ray diffraction and electron microscopy (Chaps. 3 and 11), should eventually help explain their roles in determining photosynthesis efficiency [89, 90]. Here, we provide a brief description of the structure and function of PSII. A number of extensive texts and several comprehensive review articles on these subjects are available [89, 90, 98].

PSII is a complex that catalyzes the thermodynamically most demanding photochemical reactions in living systems: absorption of light, the splitting of water into oxygen, reducing cofactors necessary to fix carbon dioxide, and the production of organic molecules [17, 42, 89, 99]. It is a unique process of water oxidation requiring absorption of four light quanta that provides almost all of the Earth's oxygen. Although some other inorganic chemical reactions can do a similar job, those reactions are so violent that biological processes would not be able to sustain them [89]. Therefore, the evolution of PSII was absolutely crucial for the emergence and endurance of green plants [95]. The PSII reaction centre complex is composed of D1 and D2 proteins binding six chlorophylls, two pheophytins, and two β-carotenes. The central cofactors, chlorophylls and pheophytins, involved in charge transfer are arranged in two quasisymmetric D1/D2 branches. The charge separation mainly proceeds along the D1 branch [100, 101]. On excitation, the energy is delocalized among these cofactors leading to collective Frenkel excitonic excited states [5, 16, 17]. These are mixed with charge-transfer states, that is, these excitons exhibit a non-uniform electron density [95, 102]. These states provide ultrafast channels for both exciton relaxation and charge transfer. However, as in LHC [61, 74, 75], the reaction centre also has to cope with energetic disorder. Fast nuclear motions (intra- and inter-pigment and protein vibrations) give rise to dynamic disorder, whereas slow protein motions constitute static disorder. This disorder determines the energy landscape (Chap. 9), which modulates the energy of the electronic states and determines the most favourable electronic configuration to initiate charge separation. Owing to the likelihood of multiple protein conformations within the ensemble, unproductive conformations may also appear, leading to energy losses. A possible mechanism to minimize such losses is to initiate charge separation from several different protein conformations [103]. Despite present knowledge about the possible pathways and timescales of charge separation, the precise mechanism responsible for the high efficiency of this process remains unknown. Here we address three key questions: (i) is electronic coherence present in charge transfer occurring in the PSII reaction centre?

(ii) does the charge transfer involve multiple pathways? and (iii) if present, does quantum coherence enhance the charge-separation efficiency?

7.5.2 *Observation of Quantum Coherence PSII Charge Transfer*

It is well-established that effective photosynthesis requires an efficient charge separation and transfer across protein complexes [104]. Yet, many aspects of how protein structure and dynamics control charge transfer, especially on the ultrafast timescale during the early stages of photosynthesis is only partly understood. In the 1990s, the remarkable direct visualization of coherent nuclear motion in the excited state of the primary electron donor in the bacterial reaction centre suggested the functional role of protein dynamics in the primary charge-transfer reaction [105]. However, no clear evidence of quantum coherence was observed. More recently, transient absorption spectra has shown that protein motions modulate the electron-transfer rate [106, 107], but the quantum coherent nature of the transfer has remained illusive. The observation of long-lived coherence in photosynthetic LHCs (Sects. 7.4 and 4.6) triggered intense experimental and theoretical efforts to establish the role of quantum coherence in promoting the efficiency of photosynthesis [19–21, 23, 34, 42, 95]. In the context of charge transfer, quantum coherence between the electronic states involved might introduce wavefunction correlations, enabling the excitation to transfer rapidly, and at the same time coherently sample multiple pathways without losing energy. Therefore, it may render the process of energy and charge transfer less sensitive to the intrinsic disorder of pigment-protein complexes and allow these systems to successfully reach their final state, avoiding energy losses [95].

When P680, the electron donor in PSII, is photoexcited, an electron transfers through a bridging chlorophyll to a pheophytin acceptor within $\sim$2 ps [106, 107]. Experimental studies of this process have focused on understanding the speed and high efficiency of the reaction: an electron is transferred for each photon absorbed [104]. Conventional electron-transfer theory fails to account for such reactions, because it assumes that electron transfer is slow compared to the relaxation of the protein complex as a whole, and the complex therefore equilibrates before the electron-transfer step. However, spectroscopic measurements have shown that the protein motion cannot keep up with the pace of charge separation [106, 107]. If so, does the protein complex dynamics play any role in the reaction centre function? Wang *et al.* suggest that the slow protein dynamics may help overcome barrier potentials produced by the membranes or by the environmental factors that could potentially slow down the electron-transfer rate [106, 107]. The influence of biomolecular dynamics on electron-transfer rates is not limited to the regime of ultrafast electron-transfer reactions. Recent studies of nonadiabatic electron transfer over long distances (Chap. 9) have shown that protein motion is crucial for electronic coupling interactions [108–110]. Other studies have also shown that the resonant electronic-vibrational configuration sustains, regenerates, or even creates coherence between electronic states during the timescale of energy and electron transfer. This mechanism does not require coherent laser excitation; it is also valid for incoherent sunlight excitation [111, 112]. To obtain a comprehensive understanding of the PSII reaction centre functions, experimental (2D spectroscopy) and theoretical methods [95], have been combined to show the electronic-vibrational (vibronic) coherences occur within reaction centre and that these coherences strongly correlate with efficient and ultrafast charge separation. Their observations demonstrated that (i) long-lived coherences are observed in the PSII reaction centre at room temperature for ~400 fs, (ii) they have vibronic origin (quantum beat frequencies match those of chlorophyll vibrational modes), (iii) they persist on the timescale of charge separation, and (iv) multiple exciton-charge-transfer pathways exist to initiate charge separation. Furthermore, observations based on the relative amplitude of quantum beats also indicate that a large fraction of the reaction centre exhibit correlated dynamics [95]. Therefore, the structure as a whole makes essential contributions to the ultimate high quantum efficiency.

Calculated population dynamics have demonstrated that the efficiency and speed of energy and electron transfer on ultrafast timescales in the PSII reaction centre strongly correlate with the degree of electronic coherence between the excited and primary charge transfer states as well as with surrounding cofactors [95]. The stronger the coherence, the faster and more efficient is the process. Thus, as evidenced by ultrafast 2D spectroscopy and theoretical modeling, the electronic coherence undoubtedly plays a major role in photosynthesis [19–21, 23, 34, 42, 95]. Therefore, in line with recent theoretical work [112–115], it seems established that vibronic coherences survive unfavorable background noise and may play an essential role in charge-separation dynamics. It seems evident that photosynthesis has evolved delicately tuned and robust "machinery" to provide the required exciton and vibrational mode matching for efficient solar-energy conversion. It operates on the principles of a quantum-designed light trap [116].

7.6 Carotenoids in Photosynthesis

Carotenoids are, along with chlorophylls, another class of pigments, that are abundant in nature. They are found in most organisms, but also can be synthesized by plants and microorganisms [22]. Although they are perhaps best known for their bright colours, they have multiple functions. They serve as light-harvesting pigments covering a region of the spectral range not accessible by chlorophylls, and they protect against excessive light by quenching excited states of chlorophylls [22, 117–120]. Other than these, they are known as efficient quenchers of highly reactive singlet oxygen and other reactive radicals by intercepting the chain of oxidative reactions [22, 121, 122]. There is accumulating evidence that this anti-oxidative function is a key mechanism of protection against various diseases, including cancer, arteriosclerosis, and macular degeneration in humans. Yet knowledge of the detailed molecular mechanism of such actions is very limited [121, 122].

The diversity of carotenoid functions is unmatched by any other class of natural pigments. The functional variety is directly related to their unique spectroscopic properties resulting from the structure of the carotenoid molecule [22]. Like in LHCs and reaction centres, an approach employing femtosecond time-resolved spectroscopy and electron microscopy could potentially provide ideal platforms for experimental and theoretical investigations of carotenoids dynamics.

References

1. E. Schrödinger, *What is Life?* (Cambridge University Press, Cambridge, 1992)
2. H.C. Longuet-Higgins, Biophys. J. **2**, 207 (1962)
3. P.C.W. Davies, *Quantum Aspects of Life* (Imperial College Press, London, 2008)
4. N. Lambert, Y.-N. Chen, Y.-C. Cheng, C.-M. Li, G.-Y. Chen, F. Nori, Nat. Phys. **9**, 10 (2013)
5. M. Mohseni, Y. Omar, G.S. Engel, M.B. Plenio (eds.), *Quantum Effects in Biology* (Cambridge University Press, Cambridge, 2014)
6. P.G. Wolynes, Proc. Nat. Acad. Sci. USA **106**, 17247 (2009)
7. S. Mukamel, Anu. Rev. Phys. Chem. **51**, 691 (2000)
8. M. Zanni, R.M. Hochstrasser, Curr. Opin. Struct. Biol. **11**, 516 (2001)
9. H. Haug, S.W. Koch, *Quantum Theory of the Optical and Electronic Properties of Semiconductors*, 4th edn. (World Scientific, Singapore, 2004)
10. R.W. Boyd, *Nonlinear Optics*, 3rd edn. (Academic Press, London, 2008)
11. M.O. Scully, M.S. Zubairy, *Quantum Optics* (Cambridge University Press, Cambridge, 2001)
12. G. Lanzani, *The Photophysics Behind Photovoltaics and Photonics* (Wiley-VCH, Weinheim, 2012)
13. T. Kobayashi, *J-aggregates* (World Scientific, Singapore, 1996)
14. G.S. Agarwal, *Quantum Optics*, vol. 70 (Springer, New York, 1974)
15. U. Akram, Z. Ficek, S. Swain, Phys. Rev. A **62**, 013413 (2000)
16. H. van Amerongen, L. Valkunas, R. van Grondelle, *Photosynthetic Excitons* (World Scientific, Singapore, 2000)
17. R.E. Blankenship, *Molecular Mechanisms of Photosynthesis*, 2nd edn. (Wiley-Blackwell, Hoboken, 2014)
18. R.J. Cogdell, A. Gall, J. Köhler, Quart. Rev. Biophys. **39**, 227 (2006)
19. G.S. Engel, T.R. Calhoun, E.L. Read, T.-K. Ahn, T. Mančal et al., Nature **446**, 782 (2007)
20. A. Ishizaki, T.R. Calhoun, G.S. Schlau-Cohen, G.R. Fleming, Phys. Chem. Chem. Phys. **12**, 7319 (2010)
21. M. Sarovar, A. Ishizaki, G.R. Fleming, K.B. Whaley, Nat. Phys. **6**, 462 (2010)
22. V. Sundström, Annu. Rev. Phys. Chem. **59**, 53 (2008)
23. H. Lee, Y.-C. Cheng, G.R. Fleming, Science **316**, 1462 (2007)
24. L.K. Grover, Phys. Rev. Lett. **79**, 325 (1997)
25. V. Vedral, Nature **453**, 1004 (2008)
26. G.R. Fleming, *Chemical Applications of Ultrafast Spectroscopy* (Oxford University Press, New York, 1986)
27. O. Svelto, S. De Silvestri, G. Denardo, *Ultrafast Processes in Spectroscopy* (Plenum Press, New York, 1996)
28. M. Braun, P. Gilch, W. Zinth, *Ultrashort Laser Pulses in Biology and Medicine* (Springer, Berlin, 2008)
29. C. Manzoni, D. Polli, G. Cerullo, Rev. Sci. Instrum. **77**, 023103 (2006)

30. D. Polli, L. Luer, G. Cerullo, Rev. Sci. Instrum. **78**, 103108 (2007)
31. P. Vasa, W. Wang, R. Pomraenke, M. Lammers, M. Maiuri et al., Nat. Photon. **7**, 128 (2013)
32. A.T. Yeh, G. Cerullo, U. Banin, A. Mews, A.P. Alivisatos, C.V. Shank, Phys. Rev. B **59**, 4973 (1999)
33. T. Brixner, J. Stenger, H.M. Vaswani, M. Cho, R.E. Blankenship, G.R. Fleming, Nature **434**, 625 (2005)
34. E. Collini, C.Y. Wong, K.E. Wilk, P.M.G. Curmi, P. Brumer, G.D. Scholes, Nature **463**, 644 (2010)
35. M.L. Cowan, J.P. Ogilvie, R.J.D. Miller, Chem. Phys. Lett. **386**, 184 (2004)
36. T. Brixner, I.V. Stiopkin, G.R. Fleming, Opt. Lett. **29**, 884 (2004)
37. T. Zhang, C.N. Borca, X.Q. Li, S.T. Cundiff, Opt. Express **13**, 7432 (2005)
38. U. Selig, F. Langhojer, F. Dimler, T. Löhrig, C. Schwarz et al., Opt. Lett. **33**, 2851 (2008)
39. D. Brida, C. Manzoni, G. Cerullo, Opt. Lett. **37**, 3027 (2012)
40. S.H. Shim, D.B. Strasfeld, Y.L. Ling, M.T. Zanni, Proc. Natl. Acad. Sci. USA **104**, 14197 (2007)
41. E.M. Grumstrup, S.H. Shim, M.A. Montgomery, N.H. Damrauer, M.T. Zanni, Opt. Express **15**, 16681 (2007)
42. K.L.M. Lewis, J.P. Ogilvie, J. Phys. Chem. Lett. **3**, 503 (2007)
43. U. Selig, C.F. Schleussner, M. Foerster, F. Langhojer, P. Nuernberger, T. Brixner, Opt. Lett. **35**, 4178 (2010)
44. B.A. West, J.M. Womick, A.M. Moran, J. Phys. Chem. A **115**, 8630 (2011)
45. S. Tanaka, S. Mukamel, Phys. Rev. Lett. **89**, 043001 (2002)
46. I.V. Schweigert, S. Mukamel, Phys. Rev. Lett. **99**, 163001 (2007)
47. G.S. Schlau-Cohen, A. Ishizaki, G.R. Fleming, Chem. Phys. **386**, 1 (2011)
48. O. Kühn, V. Sundström, J. Phys. Chem. B **101**, 3432 (1997)
49. O. Kühn, V. Sundström, J. Chem. Phys. **107**, 4154 (1997)
50. M. Dahlbom, T. Pullerits, S. Mukamel, V. Sundström, J. Phys. Chem. B **105**, 5515 (2001)
51. T. Meier, V. Chernyak, S. Mukamel, J. Phys. Chem. B **101**, 7332 (1997)
52. D. Scholes, Nat. Phys. **6**, 402 (2010)
53. A. Camara-Artigas, R.E. Blankenship, J.P. Allen, Photosynth. Res. **75**, 49 (2003)
54. J. Adolphs, T. Renger, Biophys. J. **91**, 2778 (2006)
55. T. Pullerits, M. Chachisvilis, V. Sundström, J. Phys. Chem. **100**, 10787 (1996)
56. R. Monshouwer, M. Abrahamsson, F. van Mourik, R. van Grondelle, J. Phys. Chem. B **101**, 7241 (1997)
57. G. Panitchayangkoon, D. Hayes, K.A. Fransted, J.R. Caram, E. Harel et al., Proc. Nat. Acad. Sci. USA **107**, 12766 (2010)
58. T. Meier, Y. Zhao, V. Chernyak, S. Mukamel, J. Chem. Phys. **107**, 3876 (1997)
59. Y.-C. Cheng, G.R. Fleming, Annu. Rev. Phys. Chem. **60**, 241 (2009)
60. T. Renger, Photosynth. Res. **102**, 471 (2009)
61. A. Ishizaki, G.R. Fleming, Proc. Nat. Acad. Sci. USA **106**, 17255 (2009)
62. J.M. Dawlaty, A. Ishizaki, A.K. De, G.R. Fleming, Philos. Trans. R. Soc. Lond. A **370**, 3672 (2012)
63. G.D. Scholes, T. Mirkovic, D.B. Turner, F. Fassioli, A. Buchleitner, Energy Environ. Sci. **5**, 9374 (2012)
64. A. Ishizaki, G.R. Fleming, J. Phys. Chem. B **115**, 6227 (2011)
65. C.Y. Wong, R.M. Alvey, D.B. Turner, K.E. Wilk, D.A. Bryant et al., Nat. Chem. **4**, 396 (2012)
66. T. Brixner, T. Mančal, I.V. Stiopkin, G.R. Fleming, J. Chem. Phys. **121**, 4221 (2004)
67. S. Liberato, M. Ueda, Phys. Rev. E **84**, 051122 (2011)
68. S. Hoyer, M. Sarovar, K.B. Whaley, New J. Phys. **12**, 065041 (2010)
69. F. Fassioli, A. Olaya-Castro, New J. Phys. **12**, 085006 (2010)
70. J.S. Briggs, A. Eisfeld, Phys. Rev. E **83**, 051911 (2011)
71. W.H. Miller, J. Chem. Phys. **136**, 210901 (2012)
72. M.M. Wilde, J.M. McCracken, A. Mizel, Proc. R. Soc. A **446**, 1347 (2010)
73. C.-M. Li, N. Lambert, Y.-N. Chen, G.-Y. Chen, F. Nori, Sci. Rep. **2**, 885 (2011)

74. M.B. Plenio, S.F. Huelga, New J. Phys. **10**, 113019 (2008)
75. M. Mohseni, P. Robentrost, S. Lloyd, A. Aspuru-Guzik, J. Chem. Phys. **129**, 176106 (2008)
76. H. Lee, Y. -C. Cheng, G.R. Fleming, in *Ultrafast Phenomena XVI*, ed. by P. Corkum, S. Silvestri, K. A. Nelson, E. Riedle, and, R. W. Schoenlein (Springer, Berlin, 2009) p. 607
77. Y.-C. Cheng, G.S. Engel, G.R. Fleming, Chem. Phys. **341**, 285 (2007)
78. P.K. Ghosh, A.Y. Smirnov, F. Nori, J. Chem. Phys. **134**, 244103 (2011)
79. J. Wu, F. Liu, J. Ma, R.J. Silbey, J. Cao, J. Chem. Phys. **137**, 174111 (2012)
80. V.P. Singh, M. Westberg, C. Wang, P.D. Dahlberg, T. Gellen et al., J. Chem. Phys. **142**, 212446 (2015)
81. P.D. Dahlberg, G.J. Norris, C. Wang, S. Viswanathan, V.P. Singh, G.S. Engel, J. Chem. Phys. **143**, 101101 (2015)
82. M. Schmidt am Busch, F. Müh, M. E. Madjet, T. Renger, J. Phys. Chem. Lett. **2**, 93 (2011)
83. G. Ritschel, J. Roden, W.T. Strunz, A. Aspuru-Guzik, A. Eisfeld, J. Phys. Chem. Lett. **2**, 2912 (2011)
84. C. Olbrich, T.L.C. Jansen, J. Liebers, M. Aghtar, J. Strümpfer et al., J. Phys. Chem. B **115**, 8609 (2011)
85. Y.-C. Cheng, R.J. Silbey, Phys. Rev. Lett. **96**, 028103 (2006)
86. A.K. Ringsmuth, G.J. Milburn, T.M. Stace, Nat. Phys. **8**, 562 (2012)
87. J. Dostál, T. Mančal, R. Augulis, F. Vácha, J. Pšencík, D. Zigmantas, J. Am. Chem. Soc. **134**, 11611 (2012)
88. J.M. Anderson, W.S. Chow, Y.-I. Park, Photosyn. Res. **46**, 129 (1995)
89. N. Anderson, A. Ben-Shem, Nat. Rev. Mol. Cell Biol. **5**, 971 (2004)
90. O. Einsle, Angew. Chem. Int. Ed. **53**, 7988 (2014)
91. J. Fajer, D.C. Brune, M.S. Davis, A. Forman, L.D. Spaulding, Proc. Nat. Acad. Sci. USA **12**, 4956 (1975)
92. N. Makri, E. Sim, D.E. Makarov, M. Topaler, Proc. Nat. Acad. Sci. USA **93**, 3926 (1996)
93. W.A. Cramer, G.M. Soriano, M. Ponomarev, D. Huang, H. Zhang et al., Annu. Rev. Plant Physiol. Plant Mol. Biol. **47**, 477 (1996)
94. W. Junge, Proc. Nat. Acad. Sci USA **96**, 4735 (1999)
95. E. Romero, R. Augulis, V.I. Novoderezhkin, M. Ferretti, J. Thieme et al., Nat. Phys. **10**, 676 (2014)
96. J. Nield, E.V. Orlova, E.P. Morris, B. Gowen, M. van Heel, J. Barber, Nat. Struct. Biol. **7**, 44 (2000)
97. C.W. Hoganson, G.T. Babcock, Science **277**, 1953 (1997)
98. T. Cardona, A. Sedoud, N. Cox, A.W. Rutherford, Biochim. Biophys. Acta **1817**, 26 (2012)
99. J. Barber, Quart. Rev. Biophys. **36**, 71 (2003)
100. B.A. Diner, F. Rappaport, Annu. Rev. Plant Biol. **53**, 551 (2002)
101. M.A. Steffen, K. Lao, S.G. Boxer, Science **264**, 810 (1994)
102. E. Romero, B.A. Diner, P.J. Nixon, W.J. Coleman, J.P. Dekker, R. van Grondelle, Biophys. J. **103**, 185 (2012)
103. E. Romero, I.H.M. van Stokkum, V.I. Novoderezhkin, J.P. Dekker, R. van Grondelle, Biochemistry **49**, 4300 (2010)
104. M.C. Wakeham, M.R. Jones, Biochem. Soc. Trans. **133**, 851 (2005)
105. M.H. Vos, F. Rappaport, J.-C. Lambry, J. Breton, J.-L. Martin, Nature **363**, 320 (1993)
106. S.S. Skourtis, D.N. Beratan, Science **316**, 703 (2007)
107. H. Wang, J.P. Allen, J.C. Williams, S. Blankert, C. Laser, N.W. Woodbury, Science **316**, 747 (2007)
108. T.R. Prytkova, I.V. Kurnikov, D.N. Beratan, Science **315**, 622 (2007)
109. S.S. Skourtis, I.A. Balabin, D.N. Beratan, Proc. Natl. Acad. Sci. U.S.A. **102**, 3552 (2005)
110. K.E. Wheeler, J.M. Nocek, D.A. Cull, L.A. Yatsunyk, A.C. Rosenzweig, B.M. Hoffman, J. Am. Chem. Soc. **129**, 3906 (2007)
111. F. Fassioli, A. Olaya-Castro, G.D. Scholes, J. Phys. Chem. Lett. **3**, 3136 (2012)
112. A.W. Chin, J. Prior, R. Rosenbach, F. Caycedo-Soler, S.F. Huelga, M.B. Plenio, Nat. Phys. **9**, 113 (2013)

113. N. Christensson, H.F. Kauffmann, T. Pullerits, T. Mančal, J. Phys. Chem. B **116**, 7449 (2012)
114. V. Tiwari, W.K. Peters, D.M. Jonas, Proc. Nat. Acad. Sci. USA **110**, 1203 (2013)
115. A. Kolli, E.J. O'Reilly, G.D. Scholes, A. Olaya-Castro, J. Chem. Phys. **137**, 174109 (2012)
116. R. van Grondelle, V.I. Novoderezhkin, Nature **463**, 614 (2010)
117. Y. Koyama, Y.M. Kuki, P.P. Andersson, T. Gillbro, Photochem. Photobiol. **63**, 243 (1996)
118. H.A. Frank, R.J. Cogdell, Photochem. Photobiol. **63**, 257 (1996)
119. T. Ritz, A. Damjanovic, K. Schulten, J.P. Zhang, Y. Koyama, Photosynth. Res. **66**, 125 (2000)
120. H. van Amerongen, R. van Grondelle, J. Phys. Chem. B **105**, 604 (2001)
121. J.T. Landrum, R.A. Bone, Arch. Biochem. Biophys. **385**, 28 (2001)
122. N.H. Nishino, J. Cell. Biochem. **67**, 86 (1997)

Chapter 8
Ultrafast Quantum Mechanical Processes in Animals

Abstract Several biologically-relevant functionalities are known which are initiated by ultrafast elementary photochemical reactions like energy/charge transfer, radical pair production and *cis-trans*-isomerizations. Such reactions are driven by quantum mechanical processes, which contribute to making them fast, efficient and robust. This chapter presents an overview of the current understanding of ultrafast interactions involved in primary light-initiated reactions that govern magnetoreception and vision in animals. Two other mechanisms, photoreactivation of DNA and olfaction, are also mentioned.

8.1 Introduction

Several phenomena in biological systems are explicitly quantum mechanical in nature, including, for example, the formation and breaking of chemical bonds and the accompanying molecular dynamics. Molecular properties of entities like myoglobin, ATP, and quinones—important in charge/energy transfer chains—are also known to be governed by non-classical processes [1–4]. Indeed, the biological relevance of quantum effects is being established with discoveries that there exist a class of bio-processes that are not possible to rationalize entirely on the basis of classical dynamics. The previous chapter, and the next one, discuss how the high efficiencies and adaptivity of photosynthesis and charge transfer in proteins are attributable to quantum mechanical effects—like superposition and tunneling—in the initial stages of these processes. Here, we present an overview of few of light-induced physiological processes in animals; we shall see that the quantum mechanical arrangement/rearrangement of electronic states on ultrafast timescale are an essential part of the overall dynamics.

Mechanisms like avian magnetoreception and vision, discussed here, involve remarkable sequences of events. Quantum mechanics plays an important role not only in the light-induced initial step but also in the chain of chemical reactions that follow. Other examples of quantum mechanically driven bio-photonic processes in animals include deactivation of DNA excited states to avoid photodamage, vitamin D synthesis, cell photoprotection by melanin, bioluminescence and circadian clocks [1–4].

P. Vasa and D. Mathur, *Ultrafast Biophotonics*, Biological and Medical Physics, Biomedical Engineering, DOI 10.1007/978-3-319-39614-9_8

8.2 Avian Magnetoreception

There is compelling evidence based on behavioural studies that numerous organisms, including magnetotactic bacteria, insects, amphibians, fish and birds, utilize the ability of magnetoreception for orientation and navigation using the Earth's magnetic field [4–9]. Though there is no doubt that these creatures can sense weak magnetic fields, and that magnetoreception is biologically relevant as it governs orientation, navigation as well as long-range migration, the precise underlying mechanism is not entirely understood, probably because its features vary largely from animal to animal [3]. One hypothesis to explain the mechanism responsible for magnetic field sensing in homing pigeons, for instance, is that oriented chains of clusters of magnetic iron compounds (magnetite) are present in the upper beak of these birds. They are coupled to a special class of receptors such that the mechanical torque created in response to magnetic field variations acts as a polarity-sensitive magnetic compass [3, 4, 10]. Behavioural experiments on other species, for example, European robins and some migratory birds, indicate that magnetoreception in them is akin to an inclination compass, insensitive to polarity [11]. There are indications that this intriguing property of such non-polar compasses is actually based on a photoreceptor rather than a magnetoreceptor [3, 10, 12]. It is potentially a light-initiated chemical reaction occurring in cryptochrome, which is sensitive to changes in the Earth's magnetic field [13]. Other important features of this type of evidently complex magnetic sensors include (i) strongly wavelength selective response, (ii) high sensitivity to magnetic field, (iii) adaptability to slow field changes, and (iv) disruption due to magnetic pulses and weak oscillating magnetic fields [3, 10, 12, 14–17].

The first mechanism, which involves magnetic compounds, was thought to be satisfactorily explained on the basis of classical electromagnetism [4], which is, relatively, a slow mechanism. However, contradictory reports have recently appeared that show the magnetite deposits to be a kind of macrophage which do not participate in the magnetic sense [18]. The second mechanism is ill understood and is postulated to depend on the participation of two electronic states, particularly the spin singlet (S) and triplet (T) states possessing slightly different energies [4, 19]. Invoking electron spins implies that the mechanism is entirely quantum mechanical in nature, and at least for the initial light-induced steps, it involves ultrafast dynamics. The possibility that biological systems perform a kind of spin resonance experiment to sense external stimulus as an aid in yearly migration is both fascinating and intriguing.

8.2.1 Radical Pair Mechanism

To explain the observed properties of a light-dependent biological compass, radical-pair generation has been proposed as a plausible mechanism [3, 4, 19]. It was already well known that radical pairs and different spin states can mediate magnetic field sensitive photochemical reactions via Zeeman interactions, but the required field

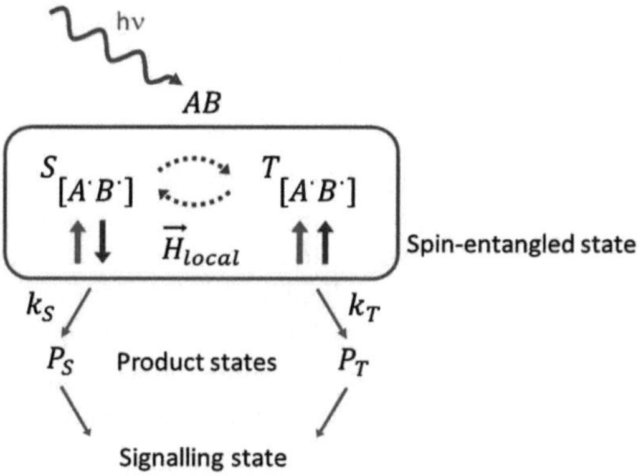

Fig. 8.1 Schematic of a radical pair scheme proposed to rationalize light-induced avian magnetoreception. The absorption of a *blue-photon* generates a radical pair $A^{\bullet}B^{\bullet}$ which exists in a coherent superposition of two spin-states, S (singlet or anti-parallel) and T (triplet or parallel). The Earth's magnetic field affects the $S \leftrightarrow T$ inter-conversion which, in turn, alters the rates (k_S and k_T) of each spin-state reaction to generate a magnetic field dependent biological response

strengths were expected to be much stronger than the Earth's magnetic field, ~45 μT. In biological systems, it has been conjectured that the quantum dynamics of non-equilibrium electron spin states of the radical pair change the yields of chemical reactions even though the Zeeman interaction with the geomagnetic field is more than six orders of magnitude smaller than the thermal energy per molecule! Classically, the effect of such a weak interaction on reaction rates would be entirely negligible [4, 7, 20]. The quantum mechanical radical pair mechanism seems to be the only model in which a weak external magnetic field is able to influence a macroscopic response via extremely feeble interactions. Although the exact details and the steps involved are quite complex, the standard radical-pair model [3, 4, 20, 21] is schematically summarized in Fig. 8.1.

A radical pair, or a pair of bound molecules with each one having an unpaired electron ($A^{\bullet}B^{\bullet}$), is created by photoinduced electron transfer. This pair is in a spin-correlated entangled state which can have either anti-parallel (S) or parallel (T) spin configuration. The superposition state then evolves under the combined effect of the Earth's magnetic field via Zeeman interactions and internal nuclear hyperfine interactions, leading to coherent inter-conversion ($S \leftrightarrow T$) between S and T states. As the inter-conversion is influenced by the local magnetic field, the lifetime of the radical pair and the fractional yields of the two products are determined by the extent and timing of the $S \leftrightarrow T$ step. Thus, the biological response becomes sensitive to the weak geomagnetic field on a microscopic level. If the radical pair has a fixed orientation or is immobilized, the tensorial nature of the interactions implies a directional magnetoreception, forming the basis of a biological photochemical compass [3, 4, 22–27].

This radical pair picture seems established and has, indeed, been successfully used over the last four decades for quantitative interpretation of a variety of experimental data [4, 28]. However, the precise nature of the radical pair that might be involved remains unknown! As magnetoreception in many birds is found to be also light sensitive, particularly to blue wavelengths, a series of radical-pair photochemical reactions occurring in cryptochromes found in the retina may be potential candidates [3, 4, 7, 29]. Some of the clearest experimental demonstrations of this fundamentally quantum-mechanical effect, without which there would be no significant response to changes in the Earth's magnetic field, are observations of quantum beats in the luminescence recombination of radical ion pairs in non-polar solvents [30–32]. Theoretically, simple models involving radical pair generation and highly anisotropic nuclear spin configurations are sufficient to explain the wavelength selectivity, directionality and the disruptive effect of time-dependent external magnetic fields on the biological compass [3, 33].

8.2.2 Photocycle of a Flavin Blue-Light Photoreceptor

Most biologically relevant photoreceptor proteins contain chromophores that perform photoinduced charge transfer, or an isomerization reaction, as the initial step of the chain of events that lead to a specific type of signalling [2, 34]. Flavin chromophores are an important class of such proteins; they mediate a variety of blue-light responses of microorganisms, plants and animals, based on photoinduced electron- and proton-transfer reactions. Flavin-binding phototropins and cryptochromes control functionalities like the direction of growth, chloroplast movement, stem elongation, leaf expansion, circadian rhythms [2, 34] and, as we have seen, avian magnetoreception [3, 4, 7, 29]. To explain the underlying response mechanism in flavin adenin dinucleotide (FAD) based blue-light receptors, reaction models involving a combination of hydrogen bond and structural rearrangements were suggested on the basis of high-resolution crystal structures [2, 35, 36]. However, ultrafast spectroscopy on two different FAD-binding proteins provided the much needed detailed insight into the reaction mechanism that leads to the signalling state stimulating the biological response [37–39]. The typical photocycle of the FAD-binding protein is schematically depicted in Fig. 8.2. Following photoabsorption by the FAD chromophore, an electron is transferred from a nearby amino acid, on picosecond timescales, to form the FAD^- anion. FAD^- is then rapidly protonated, through proton transfer from the tyrosine molecule, to form a neutral $FAD^{\bullet}Tyr^{\bullet}$ radical pair [38]. The sequence of quantum dynamical events involving proton transfers between tyrosine, glutamine, and FAD is suggested to lead to the long-lived signalling states, $FADH^{\bullet}$ and $FAD^{\bullet}Tyr^{\bullet}$, conveying the biological response: magnetoreception. Although the initial step is quite fast, the entire signalling event occurs on μs times because it takes relatively long for the spin dynamics to be affected by a very weak interaction [3, 4]. Similar photocycles, but with different time constants and intermediates, have also been postulated for other FAD-binding proteins [37, 39]. It is, therefore, believed that charge transfer processes in the flavin chromophore are the key to rationalizing the photoresponse of chryptochromes and light-induced avian magnetoreception.

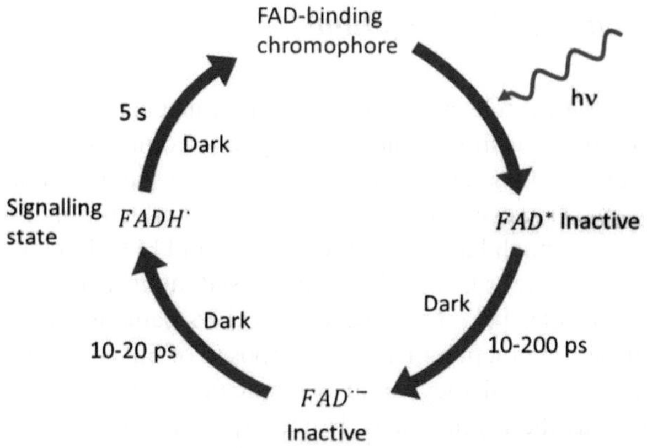

Fig. 8.2 Typical photocycle of flavin adenin dinucleotide (FAD) binding *blue-light* photoreceptors like cryptochrome. The photoinduced transient state, $FADH^{\bullet}$, formed via charge transfer reactions produces a radical pair $FAD^{\bullet}Tyr^{\bullet}$ with a nearby tyrosine molecule. The Earth's magnetic field controls the $S \leftrightarrow T$ inter-conversion of the radical pair, constituting a mechanism underlying avian magnetoreception and some other biologically relevant functionalities

8.2.3 An Effective Photochemical Compass

As interaction energies involved in magnetoreception are much smaller than thermal energies, a radical pair reaction must fulfil a number of chemical, magnetic, kinetic, structural, and dynamic requirements for the photochemical compass to be effective under ambient conditions [7]. Based on results of time-resolved spectroscopy, these can be qualitatively described as (i) the radical pair must be formed in a coherent superposition of electron-nuclear spins and each of the spin-states (S and T) should undergo a unique spin-selective reaction, (ii) availability of suitable anisotropic hyperfine interactions to enhance differences in S and T state dynamics, (iii) the radical pair lifetime must be long enough to allow the weak magnetic field to affect the spin dynamics, (iv) other spin-spin interactions, and spin relaxation mechanisms if present, should be much weaker and slower, and (v) the alignment of the radical pair must be preserved sufficiently long for directional information to be processed. These criteria are, of course, interlinked but not always mutually compatible. For example, a stronger interaction requires shorter inter-radical distances whereas weak spin-spin interaction demands sufficient separation from surrounding nuclei. Such requirements impose stringent conditions on both the protein structure and on the chemical reactions [7, 40]. These factors and their effects on spin-dynamics have been investigated experimentally as well as theoretically; however, uncertainty still prevails about the strength of the magnetic field interactions and other intrinsic as well as extrinsic relaxation mechanisms [3, 4, 26, 27, 40–43]. Detailed time-resolved structural analysis following the photoexcitation of flavin chromophore may provide more insights into the radical pair production and dynamics at various stages of the signalling chain.

8.2.4 Experimental Investigations

Apart from its biological relevance, the radical pair mechanism also offers opportunities for developing a sensitive molecular magnetic compass. Accordingly, several experimental studies of magnetic field effects on radical-pair reactions have been carried out in solutions. However, they showed detectable effects only for field strengths higher than ~10 mT, much larger than the Earth's field [44]. Recently, the observed sensitivity of such reactions has gradually increased, mainly by selection of participating organic molecules [45]. Maeda and coworkers demonstrated one of the most sensitive radical-pair interaction in a triad composed of linked carotenoid, porphyrin and fullerene molecules [46]; they found magnetic field sensitivities down to ~50 μT, of the same order as the Earth's field. However, the angular dependence could be demonstrated only for stronger field strengths, of the order of mT. Their experiment is discussed in more detail in Chap. 10 in the context of developing a biomimetic magnetic compass.

As mentioned earlier, cryptochrome is a ubiquitous photoactive protein containing FAD that resides in the retina (it is also present in plants and bacteria). It is the potential photoreceptor responsible for radical pair production [29]. Very recently, in vitro experiments performed on cryptochrome extracted from the plant *Arabidopsis thaliana* have successfully shown magnetically sensitive radical-pair reactions. Changes in reaction yield by ~25 % for a 30 mT field and ~1 % for a 1 mT field have been reported [47]. Although these are still much larger the Earth's magnetic field, it is speculated that much stronger response may be observed in vivo due to the specific arrangement of the protein and its surrounding.

The topic of enhanced magnetic interaction has also been addressed theoretically and several factors like configuration, hyperfine interaction and environmental noise have been considered [25, 48, 49]. Results indicate that the behaviour of the biological response can even be counterintuitive, for example environmental noise aiding performance (as in the case of photosynthesis) rather than affecting it adversely [50]. Other models involving long-lived dipole moments as compared to the yields of chemical reactions have also been proposed [51]. Experiments have also identified a neural substrate in pigeons which can process magnetic sense information [52]. The area of the brain that contains this substrate was shown to respond to field direction, intensity, and polarity [10, 18, 52]. More experiments on the neural properties of organisms in which the radical-pair model might be applicable are vital, and may shed further light on the relevance of the radical pair mechanism.

Experimentally, the main obstacle to verification of the radical pair model is to show that cryptochromes, or other candidate molecules, can generate radical pairs that respond to the strength as well as direction of the extremely weak geomagnetic field [7, 15, 29, 47]. Ultrafast single molecule spectroscopy and imaging techniques (Chaps. 3 and 4) may prove to be of utility in such investigations. Another intriguing aspect is that though much stronger—but slowly varying—fields do not disrupt magnetoreception, radio frequency oscillating fields as weak as 50 nT do disrupt the magnetic sense in some bird species [17, 53]. Although theories [54] have been proposed to explain these behavioural experiments [14, 15], more systematic in vivo and in vitro experiments are certainly needed to understand the mechanism. One interesting possibility is that of employing shaped magnetic pulses to coherently control the quantum dynamics to mimic the environmental response [48, 55]. Even in behavioural experiments, if it could be shown that birds are able to respond to the encoded directional information in these oscillating fields, this would strongly favour the quantum mechanical radical pair mechanism. Such effects would be almost impossible to describe using any other magnetoreception mechanism [3, 4]. Finally, although the basic principle of radical pair production seems to have been rationalized [3, 4, 20, 21], the potential chromophore has been recognized [7, 29], and the brain area active during the magnetic response has been identified [52], almost nothing is known about the signalling pathway that leads to the biological response. As cryptochrome is involved, a visual pathway seems likely [29, 33], but a deeper understanding of the chain of events connecting the S/T yield to a neurological signal is vital not only to obtain conclusive evidence in favour of the radical pair mechanism, but also to develop a comprehensive understanding of magnetoreception in general.

8.3 Ultrafast Photoisomerization in Vision

For many organisms, the ability to sense and respond to various light conditions through light-sensing proteins or photoreceptors plays a crucial role in their survival. We have seen that there are natural pigments or photoreceptors like chlorophylls, carotenoids and cryptochromes, that convert light-signals into a biological response via ultrafast quantum mechanical charge transfer reactions [2–4]. There is another class of photoreceptors that mediates biological signalling via conformational changes. An isomerization (*trans-cis* or *cis-trans*) of a carbon-carbon double bond in an extended conjugated chromophore system is one such biologically relevant conformation change [34]. These structural changes may, in turn, lead to altered interactions or binding of other components in the signalling pathway that finally triggers the biological response [56, 57]. Perhaps the best known biological photoreceptor exhibiting light-induced conformation is rhodopsin with its retinal chromophore responsible for triggering vision process in our eyes (and in other animals).

Microbial rhodopsins like bacteriorhodopsin (bR) and halorhodopsin, present in various unicellular organisms, belong to a family of photoactive proteins undergoing similar photochemical reactions but they carry out distinctly different functions of light-driven ion transport and photosensory signalling [2, 4]. Sensory rhodopsins or phototaxis receptors are the eyes of primitive organisms controlling the cells' swimming behaviour in response to changes in light intensity and colour. Even though these rhodopsin type of proteins control different signalling pathways, all of them possess the same basic configuration of seven trans-membrane helices forming the binding pocket for the retinal chromophore. Though rhodopsin and bR have been investigated most extensively, there are reports on other rhodopsin-type proteins,

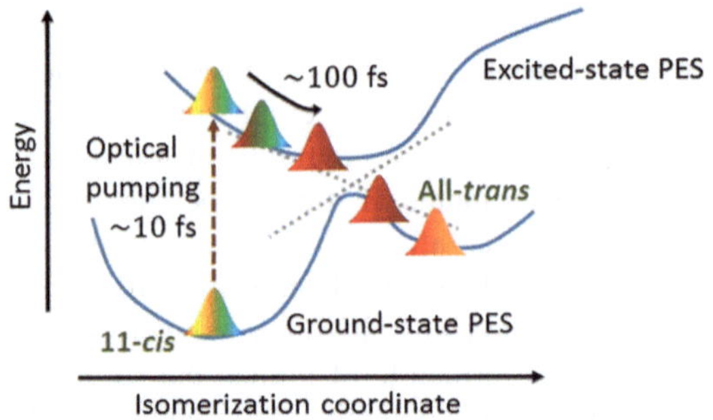

Fig. 8.3 Schematic of the isomerization potential energy surfaces (PES) of rhodopsin based on a two-state model. Ultrafast spectroscopic tracking has shown that the wavepacket dynamics through the conical intersection following impulsive excitation by a 10 fs pulse occur on timescales of 200–500 fs

confirming that these proteins are not only structurally similar but that their photocycles are also comparable [2, 4].

Absorption of light by the retinal chromophore causes it to isomerize (11-*cis* to all-*trans* for rhodopsin and all-*trans* to 13-*cis* for bR). For rhodopsin, the initial isomerization initiates a sequence of dark unidirectional reactions that ultimately leads to the detachment of the retinal chromophore from the protein [2, 4, 58]. Following early picosecond experiments on rhodopsin, ultrafast studies with improved time resolution and associated theoretical calculations identified several steps in the photocycle preceding the formation of the first stable product [59–63]. From such work, conversion of the 11-*cis* retinal chromophore to its all-*trans* form in rhodopsin was identified as the primary ultrafast (200–500 fs) event in vision [2, 4, 64]. A simple two-state model using a one-dimensional potential energy surface (PES), depicted in Fig. 8.3, was suggested to rationalize experimental observations. Accordingly, an optical excitation creates a population on the repulsive part of the excited-state PES, and the downhill motion of the population on this surface leads to the isomerization at the PE minimum (the sink, corresponding to the bond twist angle of 90°).

Ultrafast isomerization, with a high quantum yield of 0.65, and the storage of considerable energy in the first stable bathorhodopsin intermediate [65–71] suggested an unusually fast and efficient photo-activated unidirectional reaction [2, 4, 64]. Consequently, rhodopsins reactivity has been attributed to a conical intersection [72, 73] between the curves for the ground and excited electronic states (Fig. 8.3) enabling fast and efficient conversion of photon energy into a chemical signal [74–76]. The non-adiabatic coupling induced by the conical intersection is yet another example of a biologically relevant process in which quantum mechanics is expected to play an important role. However, direct experimental evidence for the involvement of a conical intersection has proved to be challenging because (i) the energy gap between the electronic states of the reacting molecule changes rapidly on ultrashort timescales, and (ii) spectroscopically, the intermediate products span a wide range of wavelengths. Therefore, the experimental technique must combine high temporal resolution (~10 fs) with a broad spectral observation window (~500–1100 nm) to follow the dynamics leading to the conical intersection in rhodopsin in real-time [64, 77].

Polli and coworkers tracked the coherent wavepacket motion of bovine rhodopsin [64] from the photoexcited Franck-Condon region to the photo-product by monitoring the loss of reactant emission and the subsequent appearance of photo-product absorption using ultrafast pump-probe spectroscopy. They also carried out molecular dynamics calculations to show excellent agreement between the observations and the expected electronic state crossing shown in Fig. 8.3, indicating that the structural evolution from the initial towards the final state is restricted almost exclusively to the central atoms in the molecule [64]. This localized deformation is promoted by the tight chromophore-protein binding that restricts other possible degrees of freedom [78]. Such a self-assembled arrangement is speculated to ensure optimum funnelling of photon energy to rapidly (within ~80 fs) drive the targeted atomic displacements. Once isomerization has occurred, the majority of highly strained structures relax rapidly in the photo-product well (all-*trans* structure), completing the primary isomerization reaction. The remaining molecules convert back to the 11-*cis* structure. These findings undoubtedly constitute compelling evidence for the existence and importance of conical intersections in ultrafast visual photochemistry. The photoisomerization reaction rate in rhodopsin approximately matches the timescale of molecular torsional vibrations involved in the process [64, 77]. Such a fast reaction is suggestive of a ballistic interaction or in-phase trajectory of the wavepacket with the molecular vibrations without any dephasing. The speculated vibrational coherences in support of the ultrafast quantum mechanical rhodopsin dynamics have been experimentally observed in real-time by employing impulsive excitation with a ~10 fs pulse. This very fast unidirectional reaction in rhodopsin is an example of the "brute-force" method applied by nature, to ensure maximum efficiency by clamping unwanted conformational deformations [2].

8.4 DNA Photoreactivation

DNA photolyases are highly efficient light-driven DNA repair enzymes, which revert the genome-damaging effects caused by ultraviolet (UV) radiation via a photoreactivation mechanism [34] in the presence of visible light (preferentially violet/blue). These enzymes occur in almost all living organisms exposed to sunlight, other than humans and other placental mammals [79, 80]. Spectroscopic and structural analysis has recently yielded a concise view of how photolyases recognize DNA lesions involving two neighbouring bases, catalyze the repair reaction within a nanosecond, and still achieve quantum efficiencies close to unity [37–39, 81, 82]. Apart from these mechanistic aspects, the potential of DNA photolyases for the generation of highly UV-resistant organisms, or for skin cancer prevention, is being increasingly recognized.

Initial experiments on photolyases mediated enzymatic photoreactivation of ultraviolet (UV)-damaged DNA were reported more than 50 years ago [81]. The repair process depends on a non-covalently bound cofactor, FAD, a redox-active catalyst, which we have seen to also participate in numerous other biological processes. Many biochemical and biophysical studies have been carried out to unravel the fundamentals of this unique mode of action. The work culminated in the elucidation of a detailed three-dimensional structure of the enzyme in 1995. However, the geometrical structure of the enzyme alone proved insufficient to fully unveil their repair action on UV-damaged DNA. Extensive spectroscopic studies have helped understand the transition-state catalytic reactions, electron transfer, bond breaking and bond forming in DNA photolyase. Recently, ultrafast studies of the photoreduction dynamics of the neutral radical flavin cofactor (FADH) to the biologically active, a fully reduced deprotonated flavin $FADH^-$ have revealed initial electron-transfer reactions, occurring on picosecond timescale [83] similar to magnetoreception. The experiments also revealed faster solvation dynamics of locally-ordered water molecules near the active site. Such dynamic solvation at binding and active sites is critical to protein recognition and enzyme catalysis [84]. As in the case of light-harvesting and charge-transfer during photosynthesis, these ultrafast ordered-water dynamics play a key role in instantaneously stabilizing $FADH^-$ to prevent fast charge recombination [83]. Undoubtedly, the combination of femtosecond spectroscopy with time-resolved structural studies will make significant contributions to the understanding of the complex DNA repair dynamics at the molecular level.

8.5 Olfactory Reception

A related area of interest, but one that does not involve photons, is olfactory reception. Olfaction is the sense of smell, or the activation of related sensory cells in invertebrates [85–87]. The mechanism underlying olfactory reception and most interestingly, how such a wide range of odours can be discriminated by a restricted number of olfactory receptors is still not understood [3, 4]. There have been studies, mainly behavioural, suggesting that odorants are not solely discriminated by their shape or the type of binding to olfactory receptors ("lock and key" model). Instead, it has been suggested that a second ingredient involving the vibrational spectrum of the odourant (and hence, possibly, ultrafast in nature) is necessary to guide the "lock and key" mechanism. It has been proposed that this second mechanism may be phonon-assisted tunneling of an electron between two receptor sites via the odourant. So far, there are contradictory reports about the role played by the vibrational contribution [88, 89]. Consequently, olfactory reception is an example where a more detailed understanding of the protein structure, as well as its interactions with a bound analyte, can elucidate the olfactory reception mechanism on a microscopic level.

References

1. B. Pullman, A. Pullman, *Quantum Biochemistry* (Wiley-Interscience, New York, 1963)
2. V. Sundström, Annu. Rev. Phys. Chem. **59**, 53 (2008)
3. N. Lambert, Y.-N. Chen, Y.-C. Cheng, C.-M. Li, G.-Y. Chen, F. Nori, Nat. Phys. **9**, 10 (2013)
4. M. Mohseni, Y. Omar, G.S. Engel, M.B. Plenio (eds.), *Quantum Effects in Biology* (Cambridge University Press, Cambridge, 2014)
5. R. Wiltschko, W. Wiltschko, *Magnetic Orientation in Animals* (Springer-Verlag, Berlin, 1995)
6. R.C. Beason, Integr. Comp. Biol. **45**, 565 (2005)
7. C.T. Rodgers, P.J. Hore, Proc. Nat. Acad. Sci. USA **106**, 353 (2009)
8. E. Wajnberg, D. Acosta-Avalos, O. Cambraia, Alves, J. Ferreira de Oliveira, R.B. Srygley, D.M. S. Esquivel, J. R. Soc. Interface **7**, S207 (2010)
9. M. Winklhofer, J.R. Soc. Interface **7**, S131 (2010)
10. R. Wiltschko, K. Stapput, P. Thalau, W. Wiltschko, J.R. Soc. Interface **7**, S163 (2010)
11. W. Wiltschko, R. Wiltschko, Science **176**, 62 (1972)
12. W. Wiltschko, R. Wiltschko, U. Munro, Naturwissenschaften **87**, 36 (2000)
13. M. Liedvogel, H. Mouritsen, J.R. Soc. Interface **7**, S147 (2010)
14. W. Wiltschko, K. Stapput, P. Thalau, R. Wiltschko, Naturwissenschaften **93**, 300 (2006)
15. T. Ritz, Procedia Chem. **3**, 262 (2011)
16. W. Wiltschko, J. Traudt, O. Güntürkün, H. Prior, R. Wiltschko, Nature **419**, 467 (2002)
17. T. Ritz, P. Thalau, J.B. Phillips, R. Wiltschko, W. Wiltschko, Nature **429**, 177 (2004)
18. C.D. Treiber, M.C. Salzer, J. Riegler, N. Edelman, C. Sugar et al., Nature **484**, 367 (2012)
19. K. Schulten, C.E. Swenberg, A. Weller, Z. Phys. Chemistry **111**, 1 (1978)
20. C.R. Timmel, K.B. Henbest, Phil. Trans. R. Soc. A **362**, 2573 (2004)
21. C.T. Rodgers, Pure Appl. Chem. **81**, 19 (2009)
22. K. Schulten, P. Tavan, Nature **272**, 85 (1978)
23. C.R. Timmel, F. Cintolesi, B. Brocklehurst, P.J. Hore, Chem. Phys. Lett. **334**, 387 (2001)
24. F. Cintolesi, T. Ritz, C.W.M. Kay, C.R. Timmel, P.J. Hore, Chem. Phys. **294**, 707 (2003)
25. J.C.S. Lau, N. Wagner-Rundell, C.T. Rodgers, N.J.B. Green, P.J. Hore, J.R. Soc. Interface **7**, S257 (2010)
26. E. Hill, T. Ritz, J.R. Soc. Interface **7**, S265 (2010)
27. I.A. Solov'yov, H. Mouritsen, K. Schulten, Biophys. J. **99**, 40 (2010)
28. K.M. Salikhov, *Magnetic Isotope Effect in Radical Reactions* (Springer, Wien, 1996)
29. T. Ritz, S. Adem, K. Schulten, Biophys. J. **78**, 707 (2000)
30. V.M. Grigoryants, B.M. Tadjikov, O.M. Usov, Y.N. Molin, Chem. Phys. Lett. **246**, 392 (1995)
31. V.A. Bagryansky, O.M. Usov, V.I. Borovkov, T.V. Kobzeva, Y.N. Molin, Chem. Phys. Lett. **255**, 237 (2000)
32. V.A. Bagryansky, V.I. Borovkov, Y.N. Molin, Russ. Chem. Rev. **76**, 493 (2007)
33. T. Ritz, M. Ahmad, H. Mouritsen, R. Wiltschko, W. Wiltschko, J.R. Soc. Interface **7**, S135 (2010)
34. M.A. van der Horst, K.J. Hellingwerf, Acc. Chem. Res. **37**, 13 (2004)
35. M. Gomelsky, G. Klug, Trends Biochem. Sci. **27**, 497 (2002)
36. W. Laan, M.A. van der Horst, I.H. van Stokkum, K.J. Hellingwerf, Photochem. Photobiol. **78**, 290 (2003)
37. M. Gauden, S. Yeremenko, W. Laan, I.H.M. van Stokkum, J.A. Ihalainen et al., Biochemistry **44**, 3653 (2005)
38. M. Gauden, I.H.M. van Stokkum, J.M. Key, D.C. Luhrs, R. van Grondelle et al., Proc. Nat. Acad. Sci. USA **103**, 10895 (2006)
39. W. Laan, M. Gauden, S. Yeremenko, R. van Grondelle, J.T.M. Kennis, K.J. Hellingwerf, Biochemistry **45**, 51 (2006)
40. I.A. Solov'yov, D. Chandler, K. Schulten, Biophys. J. **92**, 2711 (2007)
41. C.R. Timmel, U. Till, B. Brocklehurst, K.A. Mclauchlan, P.J. Hore, Mol. Phys. **95**, 71 (1998)
42. O. Effimova, P.J. Hore, Biophys. J. **94**, 1565 (2008)

43. M. Tiersch, H.J. Briegel, Phil. Trans. R. Soc. A **370**, 4517 (2012)
44. U. Steiner, T. Ulrich, Chem. Rev. **89**, 51 (1989)
45. J.R. Woodward, C.R. Timmel, K.A. McLauchlan, P.J. Hore, Phys. Rev. Lett. **87**, 077602 (2001)
46. K. Maeda, K.B. Henbest, F. Cintolesi, I. Kuprov, C.T. Rodgers et al., Nature **453**, 387 (2008)
47. K. Maeda, A.J. Robinson, K.B. Henbest, H.J. Hogben, T. Biskup et al., Proc. Nat. Acad. Sci. USA **109**, 4774 (2012)
48. J. Cai, F. Caruso, M.B. Plenio, Phys. Rev. A **85**, 040304(R) (2012)
49. E.M. Gauger, E. Rieper, J.J.L. Morton, S.C. Benjamin, V. Vedral, Phys. Rev. Lett. **106**, 040503 (2011)
50. M.B. Plenio, S.F. Huelga, New J. Phys. **10**, 113019 (2008)
51. A.M. Stoneham, E.M. Gauger, K. Porfyrakis, S.C. Benjamin, B.W. Lovett, Biophys. J. **102**, 961 (2012)
52. L.-Q. Wu, J.D. Dickman, Science **336**, 1054 (2012)
53. K. Maeda, A.J. Robinson, K.B. Henbest, H.J. Hogben, T. Biskup et al., Biophys. J. **96**, 3451 (2009)
54. J.N. Bandyopadhyay, T. Paterek, D. Kaszlikowski, Phys. Rev. Lett. **109**, 110502 (2012)
55. J. Cai, G.G. Guerreschi, H.J. Briegel, Phys. Rev. Lett. **104**, 220502 (2010)
56. K.C. Hasson, F. Gai, P.A. Anfinrud, Proc. Nat. Acad. Sci. USA **93**, 15124 (1996)
57. F. Gai, K.C. Hasson, J.C. McDonald, P.A. Anfinrud, Science **279**, 1886 (1998)
58. T. Yoshizawa, G. Wald, Nature **197**, 1279 (1963)
59. H.J. Polland, M.A. Franz, W. Zinth, W. Kaiser, E. Kolling, D. Oesterhelt, Biophys. J. **49**, 651 (1986)
60. R.A. Mathies, C.H.B. Cruz, W.T. Pollard, C.V. Shank, Science **240**, 777 (1988)
61. R.A. Mathies, S.W. Lin, J.B. Ames, W.T. Pollard, Rev. Biophys. Biophys. Chem. **20**, 491 (1991)
62. S. Roy, C.P. Singh, K.P.J. Reddy, Curr. Sci. **83**, 623 (2002)
63. S. Roy, P. Sharma, A.K. Dharmadhikari, D. Mathur, Opt. Commun. **237**, 251 (2004)
64. D. Polli, P. Altoè, O. Weingart, K.M. Spillane, C. Manzoni et al., Nature **467**, 440 (2010)
65. R.W. Schoenlein, L.A. Peteanu, R.A. Mathies, C.V. Shank, Science **254**, 412 (1991)
66. G.G. Kochendoerfer, R.A. Mathies, J. Phys. Chem. **100**, 14526 (1996)
67. H. Chosrowjan, N. Mataga, Y. Shibata, S. Tachibanaki, H. Kandori et al., J. Am. Chem. Soc. **120**, 9706 (1998)
68. G. Haran, E.A. Morlino, J. Matthes, R.H. Callender, R.M. Hochstrasser, J. Phys. Chem. A **103**, 2202 (1999)
69. H. Kandoria, Y. Furutania, S. Nishimuraa, Y. Shichidaa, H. Chosrowjanb et al., Chem. Phys. Lett. **334**, 271 (2001)
70. J.E. Kim, M.J. Tauber, R.A. Mathies, Biochemistry **40**, 13774 (2001)
71. G.A. Schick, T.M. Cooper, R.A. Holloway, L.P. Murray, R.R. Birge, Biochemistry **26**, 2556 (1987)
72. M. Garavelli, P. Celani, F. Bernardi, M.A. Robb, M. Olivucci, J. Am. Chem. Soc. **119**, 6891 (1997)
73. R. González-Luque, M. Garavelli, F. Bernardi, M. Mershin, M.A. Robb, M. Olivucci. Proc. Natl Acad. Sci. USA **97**, 9379 (2000)
74. B.G. Levine, T.M. Martinez, Annu. Rev. Phys. Chem. **58**, 613 (2007)
75. L.M. Frutos, T. Andruniow, F. Santoro, N. Ferre, M. Olivucci, Proc. Nat. Acad. Sci. USA **104**, 7764 (2007)
76. G. Tomasello, G. Olaso-González, P. Altoè, M. Stenta, L. Serrano-Andrés et al., J. Am. Chem. Soc. **131**, 5172 (2009)
77. Q. Wang, R.W. Schoenlein, L.A. Peteanu, R.A. Mathies, C.V. Shank, Science **266**, 422 (1994)
78. T. Okadaa, M. Sugiharac, A. Bondard, M. Elstnerf, P. Entelc, V. Bussg, J. Mol. Biol. **342**, 571 (2004)
79. V. Thiagarajan, M. Byrdin, A.P.M. Eker, P. Müller, K. Brettel, Proc. Nat. Acad. Sci. USA **108**, 9402 (2011)
80. C.P. Selby, S. Aziz, Proc. Nat. Acad. Sci. USA **103**, 17696 (2006)
81. S. Weber, Biochim. Biophys. Acta **1707**, 1 (2005)

82. D. Zhong, Curr. Opin. Struct. Biol. **11**, 174 (2007)
83. H. Wang, C. Saxena, D. Quan, A. Sancar, D. Zhong, J. Phys. Chem. B **109**, 1329 (2005)
84. C.-W. Chang, L. Guo, Y.-T. Kao, J. Li, C. Tan et al., Proc. Nat. Acad. Sci. USA **107**, 2914 (2010)
85. L. Turin, Chem. Senses **21**, 773 (1996)
86. J.C. Brookes, F. Hartoutsiou, A.P. Horsfield, A.M. Stoneham, Phys. Rev. Lett. **98**, 038101 (2007)
87. J.C. Brookes, Phil. Trans. R. Soc. A **368**, 3491 (2010)
88. S. Guo, J. Kim, Proteins **78**, 381 (2010)
89. M.I. Franco, L. Turin, A. Mershin, E.M.C. Skoulakis, Proc. Nat. Acad. Sci. USA **108**, 3797 (2011)

Chapter 9
Energy Landscapes, Tunneling, and Non-adiabatic Effects

Abstract This chapter introduces the important concept of energy landscapes over which protein dynamics occur. Some of such dynamics occur on ultrafast timescales via quantum mechanical tunneling and by other effects that are distinctly non-adiabatic. Both these facets are considered.

9.1 Introduction

As has been seen in earlier chapters, the focus of contemporary biologists has begun to discernibly shift from morphological explorations and phenotypic probing of organisms to seeking insights, in increasingly quantitative fashion, into the underlying mechanisms and processes at molecular levels. Thus, interconnections are now beginning to be sought between systems-level (large scale) descriptions of biological entities and processes and *molecular*-level insights. It is, therefore, quite natural to expect the physical sciences community to begin contributing to the establishment of such interconnects...and not merely as instrument providers. Apart from tools and methods that were initially developed for endeavours in the physical sciences, it is also concepts developed by physicists and chemists that are now expected to play an increasingly important role in contemporary biology. Indeed, some of these concepts have already become an accepted feature of life sciences research. In terms of tools and methods, we take note of fluorescence, Raman spectroscopy, ultra-high resolution microscopy, and imaging as prominent examples that come readily to mind. Computational modeling of single-molecule trajectories are amongst several contemporary examples of concepts that are now being modified and translated so as to make them appropriate in a life sciences scenario. For instance, protein dynamics are now modelled in terms of trajectories on energy landscapes. These landscapes are multidimensional "cousins" of conventional potential energy surfaces that physicists and chemists have long used to understand molecular dynamics, by considering the variation of potential energy as a function of internuclear coordinates. As trajectories evolve on such landscapes, processes are often discovered that can only be understood in terms of quantum mechanical effects like tunneling, and in terms

P. Vasa and D. Mathur, *Ultrafast Biophotonics*, Biological and Medical Physics, Biomedical Engineering, DOI 10.1007/978-3-319-39614-9_9

of transitions that occur in non-adiabatic fashion: all concepts that are creatively modified so as to make them applicable in the biological milieu. In the following we shall provide a working overview of these topics.

9.2 Ultrafast Processes in Proteins

It is instructive to explore the concepts of energy landscapes and dynamical processes that occur on ultrafast timescales on such landscapes—sometime involving tunneling and non-adiabatic transitions—by focusing on a specific biomolecule. The protein, myoglobin (Mb), provides an archetypal biomolecule for this purpose; it is related to the well known iron- and oxygen-binding protein in blood, hemoglobin. Mb is a protein that usually exists as a monomer; it comprises 153 amino acids that are folded into eight α-helices, each of which is interconnected to the others via short loops. Small molecules, like O_2 and CO readily bind to Mb. The resulting bond with the heme photolyzes with high efficiency, and on ultrafast times—of the order of 50 fs [1].

Results of a number of experiments, conducted using techniques like time-resolved optical absorption, transient grating spectroscopy, resonance Raman spectroscopy, circular dichroism, and photoacoustic calorimetry have offered indications that, upon irradiation by a femtosecond laser pulse, an initial ultrafast rearrangement of the heme molecule is the precursor of a whole sequence of structural changes (for a recent cogent review, see [2] and references therein). The initial ultrafast response of Mb to the severing of the Mb–CO bond, for instance, has been described in terms of a "earthquake-like" motion of Mb; the propagation through the Mb structure of the strain that is released upon ultrafast photoexcitation is said to bear similarities to how mechanical waves propagate in the course of an earthquake. Experimental indications of an earthquake-like motion have also been discovered in impulsive excitation of a bacterial photoreaction centre (Chaps. 4 and 7). Time-resolved X-ray scattering is now making it possible to track the structural dynamics of other proteins in aqueous media. For instance, it now seems established that, upon exposure to ultrafast photoexcitation, the backbone C-atoms of which protein helices are formed expand outwards from the interior; such expansion occurs on picosecond timescales [3]. Experiments on Mb recently conducted with femtosecond X-ray pulses [from the Linear Coherent Light Source (LCLS) X-ray free electron laser (FEL), as is discussed further in Chap. 11] have also helped visualization of the structural response of Mb after irradiation by a femtosecond pulse [2]. These experiments have confirmed that photo-induced perturbation at the Mb's active site propagates to the global Mb structure on ultrafast timescales, as would be expected for earthquake-like phenomena. The time evolution of the spatial extent of Mb as well as its radius of gyration show collective oscillations of the atoms in Mb, with damping that has been measured to occur on picosecond timescales [2]. Such experiments pinpoint the utility of experiments with ultrafast pulses in helping visualize the intrinsic ballistic nature of protein motion. Such motion, as indicated in Fig. 9.1, usually gets "washed out"

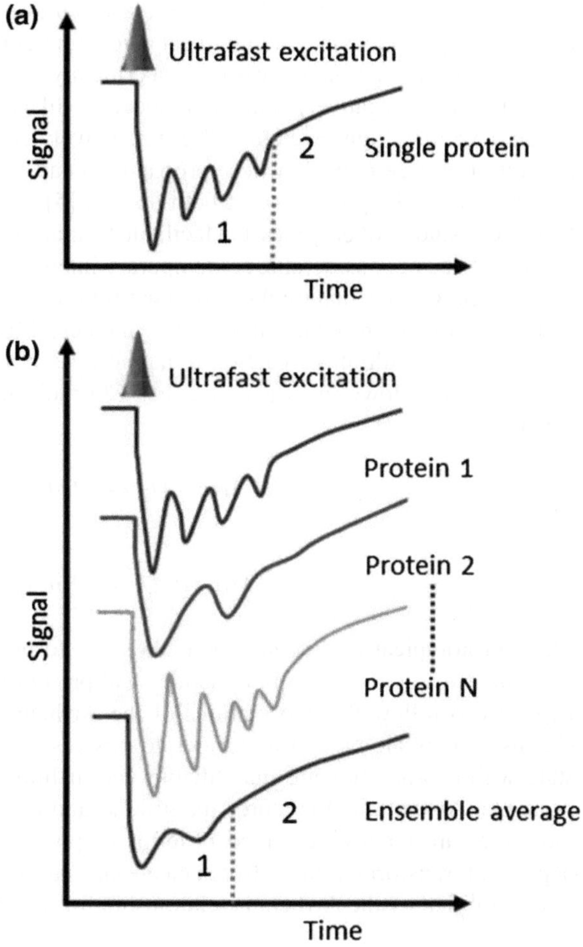

Fig. 9.1 Motions occurring within a single protein, like myoglobin, often exhibit "earthquake-like" ballistic features which are now becoming amenable to experimental investigations using ultrafast X-rays [2], from a free electron laser (Chap. 11). In an idealized single protein experiment, the time evolution of the signal being acquired will follow the behaviour shown in (**a**), with distinct oscillatory protein motion that occurs on femtosecond timescales. Analogous experiments conducted on ensembles of proteins (**b**) show "earthquake-like" transitions with femtosecond duration fine structure, readily manifesting themselves only at very short times (step 1). At longer times (step 2), thermally mediated transitions begin to dominate the dynamics, and these result in the more usual exponential-like kinetics. Note that the individual proteins in the ensemble will still exhibit the fine structure

in experiments on an ensemble of Mb molecules done on longer timescale, even though the individual proteins in the ensemble will each exhibit the signatures of the ultrafast earthquake-like ballistic events.

Nature has assigned myoglobin the important "task" of storing oxygen, to facilitate its diffusion, and to enable muscles in a human body to undergo phosphorylation. Phosphorylation is an oxidative process that drives the formation of the free-energy donor molecule, ATP. The phosphorylation process essentially involves a charge exchange reaction wherein O_2 molecules capture electrons from NADH, leading to the formation of ATP. While the physics community would be familiar with electron capture reactions occurring in single collision processes—as in interstellar space and the Earth's upper atmosphere—the capture that occurs in the life sciences environment involves the movement of electrons over considerable distances along chains of proteins. This is accomplished on ultrafast (femtosecond) timescales by traversing along protein chains via quantum mechanical tunneling [4].

The phosphorylation process plays a critically important role in metabolism, but it also produces negatively charged molecules—like excited $O_2^{-\star}$ and dianion O_2^{2-}—which have very high reactivity. These negatively charged molecules stimulate the generation of free radicals whose propagation can result in damage to cells which, in turn, can be the progenitor of both disease [5] and aging [6]. Experimental and theoretical studies of enzymes that facilitate formation of $O_2^{-\star}$ and O_2^{2-} molecular ions continue to be of considerable contemporary interest especially from the perspective of developing drugs that might lead to inhibition of their activities.

In addition to probing structural dynamics, Mb also presents several interesting facets of ultrafast reaction dynamics. As already noted, myoglobin stores O_2 molecules. However, it also attaches CO molecules. Both these processes are exemplified by

$$Mb + O_2 \leftrightarrow MbO_2, \tag{9.1}$$

$$Mb + CO \leftrightarrow MbCO. \tag{9.2}$$

These prototypical reactions, which can serve to provide an overview of some of the salient features of protein dynamics and protein function, would, of course, be expected to follow the expectations of the Arrhenius relationship. Physicists and chemists know atoms and molecules to possess a well-defined ground electronic state whose properties are quantifiable within limits imposed by the uncertainty principle. At finite temperatures the ground and excited electronic states of small molecules are readily described in terms of potential energy surfaces. The time-dependent transformations between electronic states depend on a single exponential as exemplified by the Arrhenius relationship for the rate coefficient, k_t,

$$k_t(V, T) = Ae^{-\left(\frac{V}{RT}\right)}. \tag{9.3}$$

Here, V denotes the height of the potential energy barrier, T is the temperature, R is the gas constant, and A is the pre-exponential factor. However, experiments conducted as long ago as the mid-1970s offered startling indications that even the simplest transformations involving myoglobin and other proteins did not seem to obey the well-established Arrhenius relationship. It became clear that a more sophisticated approach to reaction dynamics needed to be adopted. What prompted this realization was the observation that over the temperature range 40–200 K, the binding of O_2 or CO to myoglobin does not occur in exponential fashion. Instead, it appeared to follow a relationship that involved a *distribution* of potential barrier heights, V, with a distribution function, $\beta(V)$, denoting the probability of the barrier being at a height between the values V and $V + \delta V$.

It very quickly transpired that V needs to be expressed as a distribution rather than as a single value not just in the case of MbCO but also a host of other proteins. At both low temperatures as well as at room temperature, ultrafast protein dynamics *have* to be described in terms of distributions of V, marking a significant departure from expectations that most chemists and physicists would have from studies of simpler molecules. There are two obvious rationalizations that can be offered in order to solve this conundrum:

(i) One can postulate that all myoglobin molecules are not identical to each other. A conjecture of this type is not expected to raise alarms in the life sciences community. Biologists are, after all, accustomed to enormous degrees of non-uniformity. They are used to, for instance, observing that each of the red blood cells flowing in the reader's arteries and veins are most certainly not identical to each other. Physicists, on the other hand, would be most alarmed if they were confronted by a conjecture that the spectroscopic properties of the N_2 molecule in one part of Earth could be grossly different to those of a N_2 molecule in another part of Earth.

(ii) One could also put forward the suggestion that all myoglobin molecules may well be identical to each other but the dynamical processes they take part in proceed non-exponentially in time.

Several experiments, of different varieties, have been carried out [7], whose results unambiguously point to a clear conclusion: myoglobin molecules are *not* identical to each other. At the same time, experiments have also revealed that each myoglobin does bind to molecules like CO and O_2 exponentially.

Spectral lines of protein molecules, like Mb, exhibit inhomogeneous broadening, giving rise to a "band". Thus, it ought to be possible to make use of a sharply defined laser line to "burn" a hole into such a band. By the 1990s experimental observations of such spectral hole-burning were, indeed, being made [8]. Subsequently, it was established that not just myoglobin but proteins in general possess *distributions* of ground-state electronic levels. The energy widths of these distributions are typically measured to be larger than those in crystalline materials (typically ≤ 1 cm^{-1}) but smaller than those in amorphous materials (typically $\gg 100\,\text{cm}^{-1}$) [9]. In the case of myoglobin, such distributions of the ground state energy level cannot usually be described in terms of a single Gaussian function. In most instances, two or more Gaussian functions are required, in contrast to the prevailing situation in amorphous materials.

So, unlike the situation that prevails in small molecules, where potential energy surfaces are utilized to understand, or at least rationalize, many spectroscopic and dynamical properties, in the case of proteins it appears that there is usually a very large number of related, but distinct, conformations that constitute the "ground state". Indeed, the energy diagram is best represented not by potential energy surfaces but by energy landscapes [10]. Here, the potential energy of a given protein is expressed in terms of $(3N - 6)$ degrees of freedom. Hence, the potential energy of a protein molecule is expressed as a function that has to account for the spatial coordinates of all the N-atoms that comprise the protein: an energy landscape.

An energy landscape may be mathematically expressed, in a general sense, as a continuous function $f : X \to \mathbb{R}$, associating each conformational substate with a certain potential energy, where X denotes multi-dimensional space. We can express this mathematically by stating that for continuous space, $X = \mathbb{R}^N$. Hence, a hypersurface in $\mathbb{R}^{N+1}$ is the graphical representation of a protein's continuous energy landscape. Maxima and minima (hills and valleys) in the energy landscape represent, respectively, the local maxima and minima of f.

In order to make sense of such energy landscapes, it becomes necessary to understand the organization of the resulting hyperspace. Each valley in such hyperspace represents a specific energy substate. Within such hyperspace, each energy substate possesses a different activation barrier, thus accounting for the distributions, $\beta(V)$, discussed above. It transpires that, in practice, typical barrier heights between different valleys can have an extremely wide range of values, from ~2 meV to as high as ~1 eV.

Figure 9.2 is a schematic depiction of what a subset of a generic energy landscape of a protein might look like. The landscape can be thought of essentially as either

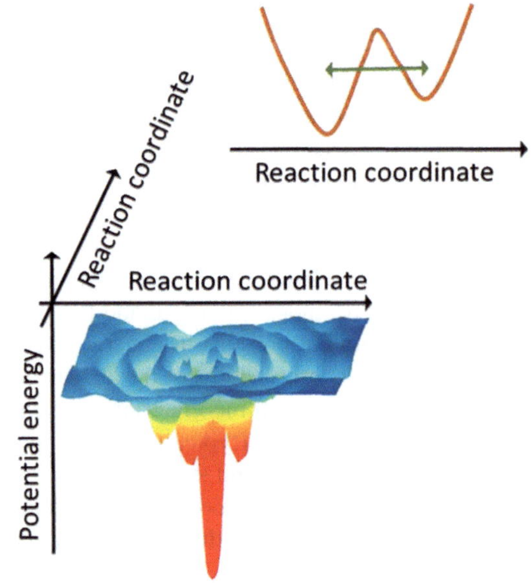

Fig. 9.2 A subset of a generic energy landscape of a protein describing the potential energy as a function of conformational or reaction coordinates. The full energy landscape would be a hypersurface in a multi-dimensional space of the coordinates of *all* atoms in the protein. At any given time, a protein is in a specific conformational substate; it constantly samples the energy landscape in order to locate conformations that are energetically more stable, with each such conformation being denoted by a minimum that possesses lower potential energy. Transformations between each subset involve tunneling between the two minima, as is indicated by the horizontal arrow in the upper figure. Such tunneling, and hence the accompanying conformational change, occurs on ultrafast timescales that can be probed using photonics techniques like two-dimensional infrared spectroscopy

a representation of all possible conformations that a protein might assume, or of the spatial positions of some of the interacting molecular entities within it and their corresponding energy levels. Such representation is of utility in, for instance, considerations of protein folding: while the structure of a protein can theoretically assume a nearly infinite number of geometrical conformations along its energy landscape, in reality proteins relax (or "fold") such that they assume a geometrical structure that has the lowest possible energy. This is represented in Fig. 9.2 as one of several minima of potential energy.

The energy landscape facilitates the description of processes that contribute to protein dynamics. In turn, the essential physics that drives protein functions can be understood in terms of energy fluctuations occurring within equilibrium—or quasi-equilibrium—conformations (corresponding to energy minima—or valleys) as well as in terms of relaxation of conformations from non-equilibrium states.

Can the structural properties of a protein be correlated to studies of relaxations and fluctuations within energy landscapes? Perhaps even more importantly, can such fluctuations and relaxations be used to determine the functional properties of a protein? Connecting protein structure with the protein's energy landscape, and relating these to protein dynamics and resulting biological function are among the key contemporary challenges in biophysics. As we shall see below, biophotonic techniques have begun to play an important role in addressing these challenges by creative adaptation of photonics techniques and methodologies.

9.3 Tunneling and Non-adiabaticity

We have already noted that experiments carried out to probe how molecules like carbon monoxide bind to myoglobin have yielded results showing that the rate coefficient for CO binding proceeds according to the expectations of the Arrhenius relationship at temperatures as low as ~20 K. At even lower temperatures, the binding rate becomes essentially independent of temperature; such temperature independence is a signature of tunneling effects.

The Arrhenius relationship contains the exponential factor $-\left(\frac{V}{RT}\right)$. Thus, one would expect that the reaction rate should vanish in the limit $T \rightarrow 0$. However, work carried out in the late 1920s by the atomic physics pioneer, Hund, offered indications that quantum mechanical effects—such as tunneling—start to assume importance at low temperatures [11]. Hängii et al. [12] have prepared a detailed exposition of tunnelling effects in biological molecules but for present purposes, it is sufficient to outline a simplified treatment for purposes of illustration. The rate coefficient k_t that quantifies the tunneling of a particle of mass M and kinetic energy E through a potential barrier, with height V and width w, can be generically written as

$$k_t = Ae^{-\left(\frac{\pi^2 w\sqrt{2ME}}{h}\right)}, \tag{9.4}$$

where h is Planck's constant. The equation indicates that tunneling occurs independently of temperature and that the propensity for tunneling decreases exponentially with distance and mass. It is now established that tunneling plays a role in biological environments in much the same manner as it does in scenarios that physical scientists are quite familiar with. It turns out that electrons, for instance, tunnel easily in biological environments; indeed, their tunneling plays a crucial role in photosynthesis, as is brought out in Chap. 7. Particles that are heavier than electrons also participate in tunneling. In the case of H^+ ions, the potential energy barrier (V) and the distance (x) through which tunneling has to occur have to assume very small values in order that tunneling effects manifest themselves. In this context, proteins again provide an excellent test-bed. The binding of CO to Mb revolves essentially around the formation of a Fe–CO bond—the Fe-atom is important as it is associated with the heme protein. At low temperatures, the potential barrier V for this bond-formation process is very small, only a few meV [13]. One would anticipate that, typically, distances involved in bond formation would be of the order of 1 nm. In line with formulation for k_t, such numbers make it entirely likely that the rate coefficient that quantifies the Fe–CO binding process would, indeed, be measurable at low values of temperatures, and be shown to be essentially independent of temperature, thereby implying the onset of tunneling.

It is instructive to probe the Fe–CO binding mechanism a little more deeply so that another important process can be unravelled: non-adiabatic transitions. In order to appreciate the meaning of adiabaticity in the context of biophotonics, it is instructive to recollect some pertinent definitions within the context of how quantum mechanics describes transitions between two distinct quantal states.

Let us consider a quantal system whose energy, at initial time t_0, is given by the Hamiltonian $\hat{H}(t_0)$. Such a system is said to be an eigenstate of $\hat{H}(t_0)$ which we label as $\psi(x, t_0)$. Let us now alter external conditions so as to change the Hamiltonian in a continuous fashion, thus giving rise to a final Hamiltonian $\hat{H}(t_1)$ at some later time t_1. The system will evolve into the final state $\psi(x, t_1)$ according to the Schrödinger equation. The concept of adiabaticity comes into play here: the alteration that we have effected in the system depends critically on the time $\tau = t_1 - t_0$ taken for the alteration to occur. In the case of a truly adiabatic process it is required that $\tau \to \infty$, wherein $\psi(x, t_1)$, the final state, becomes an eigenstate of the final Hamiltonian $\hat{H}(t_1)$, whose altered configuration can be written as $|\psi(x, t_1)|^2 \neq |\psi(x, t_0)|^2$. In order to rigourously assess whether an alteration occurs adiabatically or not, it becomes necessary to evaluate the following: (i) the energy separation between $\psi(x, t_0)$ and proximate states as well as (ii) the ratio of the interval τ to the characteristic time-scale of the evolution of $\psi(x, t_0)$ for a time-independent Hamiltonian, $\tau_{in} = \frac{2\pi\hbar}{E_0}$. Here, E_0 is the energy of $\psi(x, t_0)$. As a working definition, non-adiabatic (sometimes referred to as diabatic) and adiabatic processes can be described in the following utilitarian terms.

Non-adiabatic (diabatic) processes are those in which conditions alter so rapidly that the quantal system is unable to adapt its configuration in the course of the process. In such instances, the spatial probability density remains essentially unchanged. In other words, there is no eigenstate of the final Hamiltonian whose functional form is the same as that of the initial state. Hence, the final system can be expressed in terms of a linear combination of states that reproduce the initial probability density.

Adiabatic processes, on the other hand, are those in which conditions change sufficiently slowly, such that the quantal system is able to adapt its configuration in the course of the process. Here, the probability density does get modified by the process such that, if the quantal system starts in an eigenstate of the initial Hamiltonian, it ends in the corresponding eigenstate of the final Hamiltonian.

Consider now the situation that prior to Fe (in heme) and CO binding to each other, the lowest-energy configuration of the Fe–CO composite is a quintuplet state. Upon binding the lowest energy state must, by spin selection rules, be a singlet state. Whether or not the binding occurs depends on the magnitude of the matrix element that couples the quintuplet (q) and singlet (s) states, $\Delta = V_{qs}$. If the value of Δ approaches 0, the binding of CO to Fe will not occur and the Fe–CO composite will remain on the non-adiabatic potential energy surface that quantum mechanically describes the initial q state. On the other hand, if Δ is ascribed a sufficiently large value, the Fe–CO composite will undergo a "curve crossing" from the initial q potential energy surface to the final s surface. In the transition region, the reaction would be said to occur adiabatically, resulting in the successful binding of CO to Fe. The condition for the adiabatic crossing of potential energy curves (or surfaces) and, thereby, the deduction of the crossing probability between the q and s surfaces, can be readily deduced using the Landau-Zener formalism.

This is one more illustration that the physics that drives apparently simple atomic and molecular systems also plays a role in the much more complex world of biomolecular reactions. Consider an overview of one, highly-successful application of a biophotonic spectroscopic technique that has aided understanding of the dynamics. We invoke, once again, the attachment of CO to myoglobin.

Folded proteins are usually found to exist in multiple conformational substates, with each substate being represented by a localized minimum on the free-energy landscape. Figure 9.3 illustrates, in schematic form, a possible geometry of the Mb+CO composite along with a representation of two potential energy minima that correspond to some energy substates. Interconversion between different conformational substates of proteins is via tunneling and it has been observed to occur on very fast timescales. It has been established by means of Fourier transform IR (FTIR) spectroscopy [14] that the Mb+CO system has three absorption bands, denoted A_0 at 1965 cm^{-1}, A_1 at 1945 cm^{-1}, and A_3 at 1932 cm^{-1}. The Mb+CO composite undergoes interconversions among these three conformational substates under thermal equilibrium, with the geometrical position of the distal histidine, His-64, playing an important role in determining the energies of the three substates. Within each of the three minima in the Mb+CO energy landscape distinct fine structure is found to occur (see the inset in Fig. 9.3b) that can be discerned by means of ultrafast 2D-IR vibrational echo chemical-exchange spectroscopy [15]. Such spectroscopy has succeeded in showing how the protein's geometrical conformations switch between well-defined substates of myoglobin on timescales as short as 50 ps. The experimental signal that is measured in such studies is the growth of cross peaks (Chap. 7) in the 2D-IR spectra of CO as it binds to the heme active site. Such signals are intrinsic signatures because

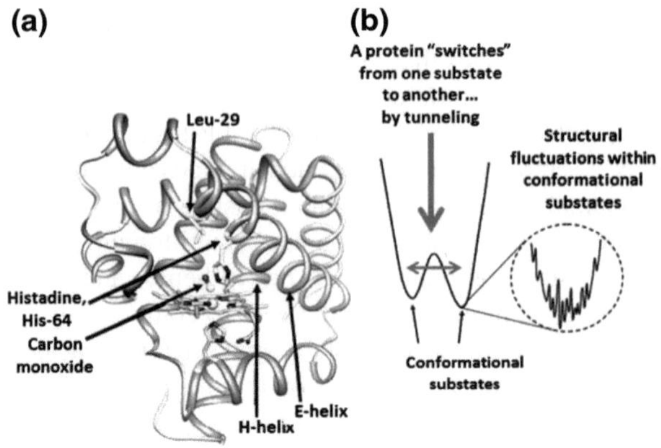

Fig. 9.3 **a** Structure of myoglobin (Mb) with a CO molecule bond to it, based on the crystal structure taken from the Protein Data Bank (ID Code 1MWC). **b** Schematic depiction of a small portion of the generic energy landscape of a protein showing two conformational subsets, as indicted by the two potential energy energy minima. Each minimum exhibits fine structure, as shown. Transformations between each subset involve tunneling between the two minima. Such tunneling, and hence the accompanying conformational change, occurs on ultrafast timescales that can be probed using photonics techniques such as 2D-IR spectroscopy

the CO-binding process involves motion of the distal histidine/E-helix that results in the location of the imidazole side group of the histidine changing. This motion is accompanied by variations in the vibrational frequency of the CO molecules which can be readily observed.

9.4 Tunneling: Dynamical Considerations

It is instructive to make some general remarks on tunneling in the context of the time it takes for a particle—or wavepacket, to be more precise—to tunnel through a barrier. Apart from its fundamental significance, the seemingly unreal phenomenon of tunneling has exerted a real and profound influence on various areas of photonics, ranging from scanning tunneling microscopy [16] to tunnel diodes [17] and, more recently, to important facets of contemporary attosecond science [18–20]. Despite this, there continues to be a lack of clarity, for instance, on exactly what tunneling time means (for a recent, cogent review, see [21] and references therein) even though experiments in the attosecond domain have begun in recent years in which tunneling time is deduced [22–25]. It appears that there are significant differences between the experimental deductions of tunneling time and those expected on the basis of established theoretical descriptions [21, 26–28]. This augers well for prospects of vigorous activity in the coming years that seek to resolve the discrepancies and, in doing so, gain deeper insights into tunneling phenomena which, in turn, will influence our understanding of tunneling within energy landscapes and in other biological situations.

In order to appreciate the challenges facing those who are bold enough to undertake work that seeks to probe tunneling phenomena in biological environments—such as tunneling on protein energy landscapes—it is instructive to consider some contemporary experimental and theoretical studies on tunneling phenomena in single atoms and molecules. We present an overview of the strategy adopted in such experiments before highlighting some of the theoretical difficulties that are expected to be encountered in attempting to extend the understanding that has been achieved of the ionization processes itself—the precursor of tunneling—in atoms to large molecules and, thence, to proteins and other biological entities.

9.4.1 Experimental Approaches

Most progress seems to have been made in the context of ionizing single atoms and molecules by exposing them to strong external fields that distort the internal Coulombic field. Femtosecond lasers have proved to be of immense use in such efforts. Commercially available lasers are readily available that enable focused intensities in the TW cm^{-2} to PW cm^{-2} range to be readily achieved in table-top laboratory systems. As has already been noted in Chap. 2, such intensities give rise to optical fields whose magnitudes match the Coulombic fields that exist within individual atoms, molecules and clusters. Irradiation of such entities by these strong optical fields inevitably results in ionization. The dynamics of ionized (ejected) electrons are highly non-linear and theoretical treatments have to be non-perturbative. The field-induced ionization process may also involve rearrangement of core electrons, not just valence shell electrons. Electrons in core orbitals necessitate consideration of dynamics that occur on attosecond timescales.

Pump-probe arrangements have usually been employed in recent experiments [23–25], with attosecond time resolution being a necessary criterion that needs to be met. Tunnel ionization induced by a strong laser field serves as the pump that starts the clock. Collision of the rescattered electron, in turn, serves as the signal (probe) that stops the clock. A somewhat different approach is to link the tunnel ionization time to the time taken by the electron to return to the vicinity of the ion core, thereby internally calibrating the attosecond clock on which all such experiments are essentially dependent. By way of illustration, we consider the latter approach [22].

The strategy is to induce tunneling by means of a probe beam that electrostatically steers, in a lateral direction, the ionized electron. Experimental implementation of this strategy has permitted measurement of the time at which the ionized electron is "born", that is, when it exits the field-distorted tunneling potential barrier, presumably with zero (thermal) instantaneous kinetic energy. As discussed in the following, the ionized electron continues to "feel" the effect of the time-dependent optical field: it oscillates to and fro with the possibility of recombining with the ionic core. On doing so, photon emission occurs, and this is the process known as high harmonic generation (see [29] and references therein). Measurement of the time that is required for both these stages—tunnel ionization and recombination—requires a perturbative probe that is able to monitor the trajectories of the ionized/rescattering electron on attosecond timescales. In [22], the task of such an ultrafast probe is ascribed to a weak second harmonic laser beam, whose optical field is described by

$$\mathbf{F}_{2\omega} = F_{2\omega}\cos(2\omega t + \phi)\widehat{\mathbf{e}_y}. \tag{9.5}$$

This second-harmonic field is kept weak and is added to the much stronger fundamental field

$$\mathbf{F}_{\omega} = F_{\omega}\cos(\omega t)\widehat{\mathbf{e}_x}, \tag{9.6}$$

that is kept orthogonally polarized to the $\mathbf{F}_{2\omega}$ field. Here, ϕ denotes the experimentally controlled temporal delay between the two fields corresponding to frequencies ω and 2ω.

The essential idea driving the experimental strategy is to rely on the weak $\mathbf{F}_{2\omega}$ field to affect only the ionized electron as it emerges from the field-distorted potential. Once the electron has tunneled through the potential barrier, the $\mathbf{F}_{2\omega}$ field causes its trajectory to be laterally displaced. This displacement, if it is large enough, would prevent the returning trajectory from encountering the ionic core, thereby "switching off" the recombination process. Hence, it becomes possible to use this lateral displacement as a gate wherein only for certain values of delay, ϕ, will the trajectories be favourable for obtaining a signal that is attributable to recombination. Measurement of high harmonic intensity as a function of ϕ thus enables determination of the exit time of the corresponding electron.

The return time can, of course, then be determined by considering the lateral velocity component of the returning electron. The weak probe beam can be thought of as an agent for breaking the time-symmetry of electron trajectories in sequential half-cycles of the optical field, $\mathbf{F}_{\omega}$, as reflected experimentally in observation of even harmonics of $\mathbf{F}_{\omega}$. The extent of symmetry breaking maximizes when the returning electron has maximum lateral velocity component at the moment of recombination. This component is amenable to experimental mapping by determining the ratio of even and odd harmonics of $\mathbf{F}_{\omega}$ as a function of ϕ, thus permitting disentanglement of the ionization and recombination times. In the case of helium, such experiments [22] have yielded ionization times that range from $\sim$300 as (attoseconds) for the 20th harmonic to as short as $\sim$100 as for the 70th harmonic. Corresponding values obtained in the case of CO_2 molecules range from $\sim$250 as for the lower (10th) harmonic to as short as $\sim$100 as for higher (30th–35th) harmonics [22].

9.4.2 *Ionization, Electron Trajectories*

There has been burgeoning interest in strong field molecular dynamics in the course of the last decade or so but, overwhelmingly, the focus of experimental attention has mainly been directed towards studying the positively-charged ions that are produced in the course of the field-molecule interaction. But high-resolution electron spectroscopy has also begun to be effectively put to use to explore the extent to which oft-used models of atomic ionization in strong fields provide an adequate enough description of *molecular* ionization phenomena. This is of particular relevance in considerations of tunneling phenomena in the biological context: can what we know about tunneling in an atomic system be applied to a large molecule and, thence, to a biological entity?

As already noted (Chap. 2), it was as long ago as 1935 that the theoretical underpinnings of the phenomenon of field-induced electron ejection from an atom were laid, when Volkov presented his exact solution of the Dirac equation in the presence of a plane-wave field [30]. Other pioneering work carried out in the 1950s and 1960s helped to substantiate the theoretical foundations of contemporary studies of field-induced ionization dynamics in strong fields, with the core of the subject relying on the theoretical treatment of electron dynamics in intense electromagnetic fields [31–35]. Contemporary descriptions of how atoms ionize upon irradiation by strong optical fields overwhelmingly rely on the Keldysh-Faisal-Reiss (KFR) model [26, 36, 37] (in conjunction with a Coulomb correction [38]) and the Ammosov-Delone-Krainov (ADK) tunneling theory [39]. In all these models, the theoretical description of the interaction of the optical field with the irradiated atom is approximated by a zero-range potential. In the case of multi-atom species, however, the spatial extent of a molecule and, for most biological entities, the associated spatial anisotropy, imply that atomic descriptions of strong-field ionization might prove totally inadequate in properly accounting for experimentally determined rates of field-induced ionization.

Over the years some attempts have been made to modify the KFR and ADK models so as to account for the effects of molecular symmetry, explicitly to account for the *shapes* of outermost molecular orbitals [40]. Experimental manifestations of such orbital effects have also been reported [41, 42]. In the case of small molecules, like N_2, reasonably good agreement has been shown to exist between experimentally measured rates of ionization and those predicted by KFR-ADK descriptions [43] as well as by modified theoretical models [44]. In the case of somewhat larger molecules (like benzene and ethylene), strong spatial alignment effects have been discerned in measured rates for total ionization and in angular distributions of ionized electrons [45] which have been amenable to interpretation in terms of the symmetry of the highest occupied molecular orbital (HOMO) for each molecule. It emerges from such studies that delocalization of the electronic charge cloud within such molecules needs to be taken into account for proper characterization of the light-molecule interaction. To this end, an appropriately length-modified Keldysh parameter (γ) [46], has been proposed, which seeks to delineate the relative importance of multiphoton ionization (MPI) and tunneling ionization (TI) in organic polyatomics. Other modifications to γ that have been suggested include structure-based [47] alterations as well as molecular orbital based alterations [48]. Figure 9.4 schematically summarizes such modifications.

As already noted, the ionization dynamics of an electron from an atom due to its irradiation by strong optical fields is theoretically described in terms of either MPI

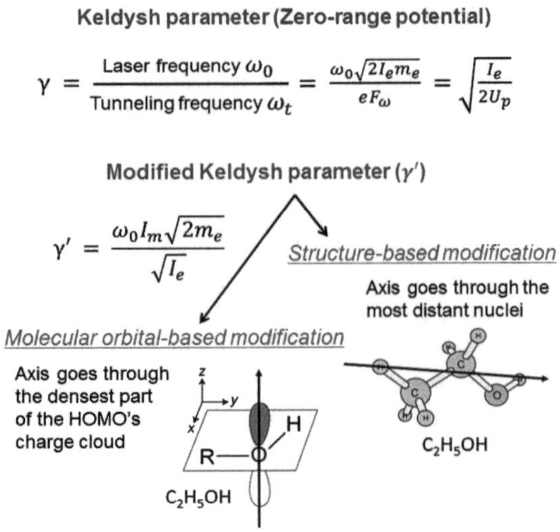

Fig. 9.4 Schematic depiction of various modifications to the KFR-ADK descriptions of ionization processes induced by strong optical fields (see text). Here, m_e denotes the electronic mass, I_m is the modified ionization energy and HOMO represents the highest occupied molecular orbital

or TI. The boundary between these two ionization regimes is expressed in terms of the adiabaticity—or Keldysh—parameter:

$$\gamma = \sqrt{\frac{I_e}{2U_p}}, \tag{9.7}$$

which can be regarded simply as the ratio of the tunneling time to the half-optical period. Here, I_e denotes the first ionization energy of the atom while U_p is the ponderomotive potential associated with the oscillating optical field. Experimentally, variation of the laser intensity (or the nature of the atom) enables a range of γ-values to be accessed and it now seems very well established that TI dominates the electron ejection dynamics for small values of γ ($\ll 1$). Under these circumstances the shape of the functional dependence of the energy of ionized electron is essentially determined by the classical propagation of the electron in the oscillating optical field. On the other hand, when γ assumes larger values ($\gg 1$), it is the multiphoton picture that more adequately describes the ionization dynamics. Here, atomic resonances and peaks in electron energy spectra that are due to above threshold ionization (ATI) manifest themselves. It is probably fair to state that a reasonably robust theoretical framework is now to hand which can be utilized to understand atomic ionization in the $\gamma \ll 1$ tunneling regime. However, corresponding theoretical clarity vis-a-vis ionization dynamics in the multiphoton regime ($\gamma \gg 1$) is still lacking.

In order to deal with ionization dynamics in molecules (as opposed to atoms), a simple working recipe has been put forward [46] that utilizes a modified, structure-based γ-parameter in order to obviate the shortcomings of a zero-range potential that is implicit in Keldysh's original γ. A similar approach might need to be adopted to extend studies to molecules that are of biological relevance, including proteins. The working recipe for small molecules comprises the following steps:

(i) Quantum chemistry is employed to carry out an ab initio geometry optimization for the neutral molecule of interest in its ground electronic state.
(ii) This geometry is then used to compute the wavefunction of the singly charged molecular ion.
(iii) On the basis of this wavefunction, an approximate rectangular potential well is derived whose width corresponds to the distance between the classical turning points within the potential; the depth of this rectangular well is made equal to the lowest ionization energy of the neutral molecule.
(iv) Superimposed on this rectangular potential is the external, time-independent potential that approximates the optical field that the molecule is irradiated with.
(v) The semi-classical Wentzel-Kramers-Brillouin (WKB) method is then utilized to deduce the tunneling rate from this field-dressed potential and the resulting tunneling frequency is used to derive a structure-modified Keldysh parameter.

Length-based γ-parameters have been applied to aromatic molecules like benzene, naphthalene and anthracene, and it has been deduced [46] that tunnel ionization requires much lower threshold values of optical field for molecules that possess spatially extended electronic orbitals in comparison to the threshold required for atoms of similar ionization energy. However, the situation is not totally clear as other experiments conducted on a different set of polyatomic molecules [49, 50] have yielded results that suggest the opposite: it is energetically more difficult to tunnel ionize some polyatomic molecules than atoms possessing similar values of ionization energy! It is likely that cognizance has to be taken of how polarizable a molecule is when making comparisons with atoms of similar ionization energy. The field-induced polarization of delocalized electrons within some polyatomic molecules would result in dynamic shielding effects as far as external fields are concerned. Such shielding may be expected to contribute additively to the intrinsic potential barrier, thus making the escape of the active electron by tunneling through the barrier energetically more demanding. Consequently, distinctly higher energy thresholds would occur for tunnel ionization of such molecules [51].

Inadequacies of a zero-range potential are a hurdle when one wishes to extend theoretical considerations from atoms to molecular systems. But there may well be another hurdle in extending an atomic theory to molecular ionization. Theoretical approaches to atomic ionization in strong fields rely on a "single active electron" picture: the applied field interacts with only one electron at a time. While this might be adequate in dealing with atoms (it is probably also inadequate when one considers multielectron atoms, like Xe), the electron dynamics that can occur *within* a molecular potential may possibly present additional problems for theorists. In this context, note that the limit $\gamma \ll 1$ that defines the tunneling regime relates only to the tunneling of a single electron though the potential barrier: it does not identify the adiabaticity of the electronic response of a molecule.

Consider, by way of illustration, field ionization of alcohol molecules. These are typical of linear molecules comprising C–H, C–O, and O–H bonds. Electrons that are associated with C–H bonds are known to be relatively unpolarizable. If one takes alcohol molecules to be alkyl derivatives of water (the H-atom in water being replaced by one or more CH_3), we can take our cue from the electronic structure of the ground electronic state of H_2O: $(1a_1)^2(2a_1)^2(1b_2)^2(3a_1)^2(1b_1)^2$. The highest occupied water orbital, $1b_1$, is almost a nearly pure $2p$ atomic orbital on oxygen and oriented perpendicular to the molecular plane. It has nonbonding character. In the case of alcohols, the highest occupied orbital retains its nonbonding character, but is now found to be somewhat mixed with the π-orbitals of the alkyl group. From the perspective of probing the efficacy, or otherwise, of the atomic KFR-ADK picture in accounting for strong-field molecular ionization dynamics, we note that alcohol molecules may simply be regarded as archetypal linear polyatomics of dimensions that are modest vis-a-vis biological molecules.

By measuring the kinetic energies of the electrons ionized upon irradiation with intense 800 nm laser light [52], it has been found, somewhat unexpectedly, that the electron ejection dynamics could be readily rationalized in terms of MPI and TI by employing the conventional (unmodified) Keldysh parameter. Linear alcohols, it would appear on the basis of electron spectroscopy results, seem to behave in atom-like fashion even though they have a much extended spatial extent. Even more unexpectedly, use of the length-modified Keldysh parameter yielded information that was inconsistent with these electron spectroscopy results insofar as disentangling the dynamics in the MPI and TI regimes was concerned [52]. These measurements have been extended by using intense pulses of 400 nm light. Here, the ponderomotive potential experienced by the ionized electron is less than the energy of a single photon and this aids the probing of the purely molecular facets of strong field dynamics [53]. However, here too it was discovered that the "atomic" Keldysh parameter suffices in adequately describing the ionization dynamics. As in the case of 800 nm light, here too the use of a length-modified γ actually yielded inconsistencies with the observed electron spectra.

It appears probable that electron spectroscopy will be a useful tool to distinguish between and discern the transition between MPI and TI. We briefly discuss the experimental signatures that are of primary importance. We do so with respect to electron spectra measured with ethanol molecules.

We show in Fig. 9.5 some typical kinetic energy spectra of ionized electrons that result upon irradiation of ethanol molecules by intense, 100 fs long pulses of 800 nm light. By varying the peak laser intensity over the range 10–200 TW cm^{-2} it becomes possible to span the range of γ-values from 2 to 0.8. The most prominent feature that stands out in these spectra is the peak structure attributable to ATI. Each of the peaks in all the spectra are separated from its neighbouring peak by a single photon energy—1.55 eV for 800 nm light. Of course these peaks originate from the multiphoton ejection of an electron from the highest occupied molecular orbital (HOMO) of each alcohol molecule, leaving the alcohol in its singly-charged ionic form. Thus, in the case of ethanol, it requires a minimum of 8 photons for ionization to occur. Additionally, the electron spectrum for ethanol also shows 9-, 10- and 11-photon ATI peaks, with similar behaviour being seen in all the other alcohols for low values of γ. As values of γ decrease, the sharp ATI peaks become

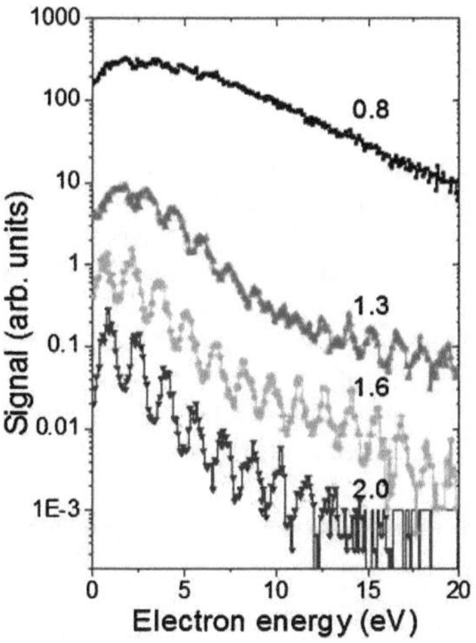

Fig. 9.5 Kinetic energy spectra of electrons ejected from ethanol (C_2H_5OH) upon irradiation by 100 fs long pulses of 800 nm light. The numbers above the individual spectra indicate the value of γ, the unmodified Keldysh parameter. γ-values could be tuned by varying the incident peak laser intensity (I) such that, for $\gamma = 0.8$, $I = 140\,\text{TW}\,\text{cm}^{-2}$ while for $\gamma = 2.0$, $I = 23\,\text{TW}\,\text{cm}^{-2}$. The structure observed at larger values of γ (in the multiphoton ionization regime) are peaks due to ATI—above threshold ionization—wherein the ejected electron continues to absorb individual 800 nm photons even when it is in the ionization continuum. This structure is "washed" out as the tunneling regime is approached. Similar behaviour has also been seen in other alcohols [53]

less pronounced, finally giving way to featureless spectra as γ approaches unity when, according to Keldysh's original formulation, ionization is essentially by tunneling. This trend validates the conjecture [52] that the transition from structured electron kinetic energy spectrum into a featureless one might be regarded a good signature of the ionization regime accessed for a given set of experimental conditions (like laser wavelength, peak laser intensity, and the zero-field ionization energy of the irradiated species). The overall morphology of each of the electron spectra shown in Fig. 9.5 is, indeed, in consonance with what one would expect from the conventional Keldysh picture and the KFR-ADK model, without recourse to any length-based or structure-based modifications.

This certainly augers well as far as future extensions of these approaches to proteins and other biological entities are concerned.

References

1. J.W. Petrich, C. Poyart, J.L. Martin, Biochemistry **27**, 4049 (1988)
2. M. Levantino, G. Schirò, H.T. Lemke, G. Cottone, J.M. Glownia et al., Nat. Commun. **6**, 6772 (2015)
3. D. Arnlund, L.C. Johansson, C. Wickstrand, A. Barty, G.J.D. Williams et al., Nat. Methods **11**, 923 (2014)

4. D. Leys, N.S. Scrutton, Curr. Opinion Struc. Biol. **14**, 642 (2004)
5. M. Valko, D. Leibfritz, J. Moncol, M.T. Cronin, M. Mazur, J. Telser, Int. J. Biochem. Cell Biol. **39**, 44 (2007)
6. S.I. Rattan, Free Radic. Res. **40**, 1230 (2006)
7. G.U. Nienhaus, R.D. Young, Encyclopedia Appl. Phys. **15**, 163 (1996)
8. J. Friedrich, Methods Enzymol. **246**, 226 (1995)
9. J.M. Vanderkooi, P.J. Angiolillo, M. Laberge, Methods Enzymol. **278**, 71 (1997)
10. H. Frauenfelder, A.R. Bishop, A. Garcia, A. Perelson, P. Schuster et al., Physica D **107**, 117 (1997)
11. V.I. Goldanskii, L.I. Trakhtenberg, V.N. Fleurov, *Tunneling Phenomena in Chemical Physics* (Gordon and Breach, New York, 1989)
12. P. Hänggi, P. Talkner, M. Borkovec, Rev. Mod. Phys. **62**, 251 (1990)
13. R.H. Austin, K.W. Beeson, L. Eisenstein, H. Frauenfelder, I.C. Gunsalus, Biochemistry **13**, 5355 (1975)
14. T.S. Li, M.L. Quillin, G.N. Phillips Jr., J.S. Olson, Biochemistry **33**, 1433 (1994)
15. H. Ishikawa, K. Kwak, J.K. Chung, S. Kim, M.D. Fayer, Proc. Natl. Acad. Sci. USA **105**, 8619 (2008)
16. G. Binnig, H. Rohrer, C. Gerber, E. Weibel, Phys. Rev. Lett. **49**, 57 (1982)
17. L. Easki, Phys. Rev. **109**, 603 (1958)
18. S. Baker, J.S. Robinson, C.A. Haworth, H. Teng, R.A. Smith et al., Science **312**, 424 (2006)
19. M. Meckel, D. Comtois, D. Zeidler, A. Staudte, D. Pavicic et al., Science **320**, 1478 (2008)
20. H.J. Wörner, J.B. Bertrand, D.V. Kartashov, P.B. Corkum, D.M. Villeneuve, Nature **466**, 604 (2010)
21. G. Orlando, C.R. McDonald, N.H. Protik, G. Vampa, T. Brabec, J. Phys. B **47**, 204002 (2014)
22. D. Shafir, H. Soifer, B.D. Bruner, M. Dagan, Y. Mairesse et al., Nature **485**, 343 (2012)
23. L. Arissian, C. Smeenk, F. Turner, C. Trallero, A.V. Sokolov et al., Phys. Rev. Lett. **105**, 133002 (2010)
24. P. Eckle, A.N. Pfeiffer, C. Cirelli, A. Staudte, R. Dörner et al., Science **322**, 1525 (2008)
25. P. Eckle, M. Smolarski, P. Schlup, J. Biegert, A. Staudte et al., Nature Phys. **4**, 456 (2008)
26. L.V. Keldysh, Sov. Phys. JETP **20**, 1307 (1965)
27. R. Landauer, T. Martin, Rev. Mod. Phys. **66**, 217 (1994)
28. Y. Ban, E.Y. Sherman, J.G. Muga, M. Büttiker, Phys. Rev. A **82**, 062121 (2010)
29. F. Krausz, M. Ivanov, M. Yu, Rev. Mod. Phys. **81**, 163 (2009)
30. D.M. Volkov, Z. Physik. **94**, 250 (1935)
31. N.D. Sengupta, Bull. Math. Soc. (Calcutta) **44**, 175 (1952)
32. Vachaspati, Phys. Rev. **128**, 664 (1962)
33. Z. Fried, Phys. Lett. **3**, 349 (1963)
34. A.I. Nishishov, V.I. Ritus, Sov. Phys. JETP **19**, 529 (1964)
35. H.R. Reiss, J. Math. Phys. **3**, 59 (1962)
36. F.H.M. Faisal, J. Phys. B **6**, L89 (1973)
37. H.R. Reiss, Phys. Rev. A **22**, 1786 (1980)
38. V.P. Krainov, J. Opt. Soc. Am. B **14**, 425 (1997)
39. M.V. Ammosov, N.B. Delone, V.P. Krainov, Sov. Phys. JETP **64**, 1191 (1986)
40. J. Muth-Bohm, A. Becker, S.L. Chin, F.H.M. Faisal, Chem. Phys. Lett. **337**, 313 (2001)
41. M. Okunishi, K. Shimada, G. Prümper, D. Mathur, K. Ueda, J. Chem. Phys. **127**, 064310 (2007)
42. D. Mathur, A.K. Dharmadhikari, F.A. Rajgara, J.A. Dharmadhikari, Phys. Rev. A. **78**, 013405 (2008)
43. T.D.G. Walsh, F.A. Ilkov, J.E. Decker, S.L. Chin, J. Phys. B **27**, 3767 (1994)
44. I.V. Litvinyuk, K.F. Lee, P.W. Dooley, D.M. Rayner, D.M. Villeneuve, P.B. Corkum, Phys. Rev. Lett. **90**, 233003 (2003)
45. T.K. Kjeldsen, C.Z. Bisgaard, L.B. Madsen, H. Stapelfeldt, Phys. Rev. A **68**, 063407 (2003)
46. M.J. DeWitt, R.J. Levis, J. Chem. Phys. **108**, 7739 (1998)
47. M.J. DeWitt, R.J. Levis, J. Chem. Phys. **110**, 11368 (1999)
48. C. Wu, Y. Xiong, N. Ji, Y. He, Z. Gao, F. Kong, J. Phys. Chem. A **105**, 374 (2001)

49. J.L. Chaloupka, J. Rudati, R. Lafon, P. Agostini, K.C. Kulander, L.F. DiMauro, Phys. Rev. Lett. **90**, 033002 (2003)
50. S.M. Hankin, D.M. Villeneuve, P.B. Corkum, D.M. Rayner, Phys. Rev. A **64**, 013405 (2001)
51. M. Lezius, V. Blanchet, M.Yu. Ivanov, A. Stolow, J. Chem. Phys. **117**, 1575 (2002)
52. T. Hatamoto, M. Okunishi, T. Lischke, G. Prümper, K. Shimada et al., Chem. Phys. Lett. **439**, 296 (2007)
53. D. Mathur, T. Hatamoto, M. Okunishi, T. Lischke, G. Prümper et al., J. Phys. Chem. A **111**, 9299 (2007)

Chapter 10
Mimicking Ultrafast Biological Systems

Abstract This chapter explores ultrafast photonic functionalities of certain proteins in plants and higher organisms and their potential role in designing new devices that mimic natural systems possessing quantum-enhanced efficiency and adaptability. The importance of quantum effects in these functionalities has recently been revealed by ultrafast spectroscopy. We discuss the role of ultrafast biophotonics in developing biomimetic devices whose quantum properties can be "engineered" for applications in light-harvesting, solar energy conversion, magnetic field sensing, photonic devices and single-biomolecular electronics.

10.1 Introduction

Evolutionary pressures have ensured that all living organisms evolve increasingly well-adapted structures. Such structures, whose development has occurred over geological timescales, are beginning to inspire new technologies; they have also started to enable developments in the emerging fields of biomimetics and bionics. These are areas in which biology-inspired solutions are sought at nano- as well as macro-scales to solve engineering challenges, such as development of systems and processes that possess high-efficiency and environmental adaptability along with self-regulating and self-healing abilities [1]. Such a biomimetic perspective requires the development of an approach to fabricate artificial systems that possess the capability to mimic all or part of a biological system [1]. Bio-inspired displays (photonic crystals mimicking butterfly wings), adhesives (velcro tapes mimicking multiply hooked structures of seeds and fruits), and water repellant surfaces (microstructures mimicking lotus leaves and petals) are amongst already-known examples of biomimetic technologies.

It is now well established that quantum effects can lead to enhanced performance, unattainable by purely classical systems, in tasks such as metrology [2], computing [3], and communication [4]. As discussed in earlier chapters (Chaps. 7, 8 and 9), there is now a compelling body of experimental evidence to suggest that quantum

P. Vasa and D. Mathur, *Ultrafast Biophotonics*, Biological and Medical Physics, Biomedical Engineering, DOI 10.1007/978-3-319-39614-9_10

processes are also critical to understanding the functional dynamics of some biological systems [5–10]. What opportunities do these observations and understanding offer for the development of ultrafast artificial systems that succeed in mimicking their natural counterparts? An important question in ultrafast biophotonics, thus, is whether it is possible to incorporate quantum effects into bio-inspired synthetic systems in order to develop devices that possess quantum-enhanced functionality. Amongst several intriguing future directions for biomimetic technologies that rely on an understanding of quantum effects in ultrafast biophotonics are light-harvesting, energy transfer, information processing, single-biomolecular devices, and several sensing applications of, for example, light and weak magnetic fields. Exploration of such applications and the development of superior materials/devices holds out the concomitant promise of achieving greater understanding, at a fundamental level, of the role of quantum mechanics in ultrafast biophotonics.

In this chapter, we focus on the opportunities presented to biomimetic technologies by natural light-harvesting, charge transfer as well as radical pair production and tunneling mechanisms discussed in Chaps. 7, 8 and 9. Some other directions, like all-optical switching and single-biomolecular electronics, are also included. We briefly present motivations for pursuing these directions, discuss some of the potential candidate architectures, and challenges that need to be overcome in order to be able to realize such applications.

10.2 Biomimetic Technologies for Novel Light-Harvesting

Most of the existing, artificial photovoltaic devices are based on large hybrid material domains that serve both as light-harvesting component as well as for transporting charge carriers [11–13]. A reason for this is that although remarkable and complex chemical architectures can now be synthesized, their assembly into macroscopic structures remains rather poorly and imprecisely controlled. On the other hand, natural systems addressing similar challenges are generally precisely self-assembled, highly efficient and remarkably photostable. As discussed in Chap. 7, the operation of the natural photosynthetic apparatus is fascinating owing to its usually complex but precise structure and the physical principles underlying its function [5–10]. Various protein units perform specialized and synchronized tasks, such as light-harvesting, short- and long-range charge separation, extraction of electrons from water, and so on. These processes are regulated by robust environment-dependent feedback and control loops [8, 14]. A key challenge in biomemetics, thus, is to learn how to construct artificial molecular devices that enable the harvesting of sunlight, and the use of such devices either for direct electric power generation (the photovoltaic approach) or to drive fuel-producing photochemical reactions (the photosynthetic approach) [11–13]. The recent observation of room temperature electronic quantum coherence, that is, the oscillatory motion of delocalized electronic wavefuctions and associated theoretical development, induced a paradigm shift in understanding the primary energy and charge-transfer processes in photosynthetic bacteria and in higher plants [5–10]. A fascinating, but so far unexplored, perspective would be to exploit such quantum coherence effects in artificial photosynthetic/photovoltaic systems to improve their performance.

Significant research efforts are presently aimed at developing artificial light-harvesting complexes (LHCs) consisting of organized assemblies of pigments and proteins; these constitute an alternative to conventional solar cell technologies [9, 15–19]. Within the framework of such tailored environments, quantum coherence could provide an additional functionality in facilitating energy transport through complex pigment-protein structures [20, 21]. This can be explored by systematic investigations of various pigment-protein configurations, with controlled distances and relative orientation, which govern the coupling between individual constituents, a primary factor in determining the extent of quantum coherence. However, fabrication of a complete artificial device, for example, a solar cell, also requires extremely challenging integration of other components, like catalysts and redox reaction steps leading to energy conversion. More diverse and feasible directions open up when attention is focused only on mimicking the enhanced light-harvesting or charge-separation functionalities of natural LHCs rather than a complete photosynthetic system [9, 15–19].

One promising direction is to mimic the extraordinary ability of some natural light-harvesting antenna complexes, such as those in green sulphur bacteria, to function under very low light conditions prevailing in deep oceans, where there is an average photon flux of only one photon every 8 h [22]. Apart from possible integration into an efficient photovoltaic device, such an artificial assembly might also work as a single-photon detector with ultra-high sensitivity [9]. Along with the enhanced energy absorption and conversion functionalities, natural LHCs also have several defence mechanisms to regulate energy absorption and transfer by protecting against photo- and/or oxidation-damage [23]. The current organic photovoltaics sorely lack such regulatory and protective mechanisms, limiting their widespread usage [24]. Hence, mimicking the regulatory mechanisms of natural complexes can lead to several unexplored possibilities in developing stable organic photovoltaics. Another potential direction to pursue is to extend the spectral range of operation of biomimetic systems from visible to far-infrared wavelengths, by incorporating different pigments [17]. Even though established inorganic and/or organic semiconductor technologies and devices for low light detection and energy conversion are already commercially available, a biomimetic light-harvesting devices can offer several advantages in terms of sensitivity, broadband ultrafast response, photostability, room temperature operation with low noise and high quantum efficiency, smaller footprint, environmental friendliness and reduced cost [9]. Therefore, over the past two decades, considerable effort has been focused on developing various biomimetic technologies (schematically depicted in Fig. 10.1) possessing 100 % efficiency in energy transfer, charge separation and photostability [19]. However, to date, no approach has been successful in integrating all these properties in a single device. This is not surprising because mimicking the entire natural photosynthetic system is a formidable task. From light capture to conversion, photosynthesis involves numerous delicately balanced reactions and excited states in proteins that are vulnerable to heat dissipation.

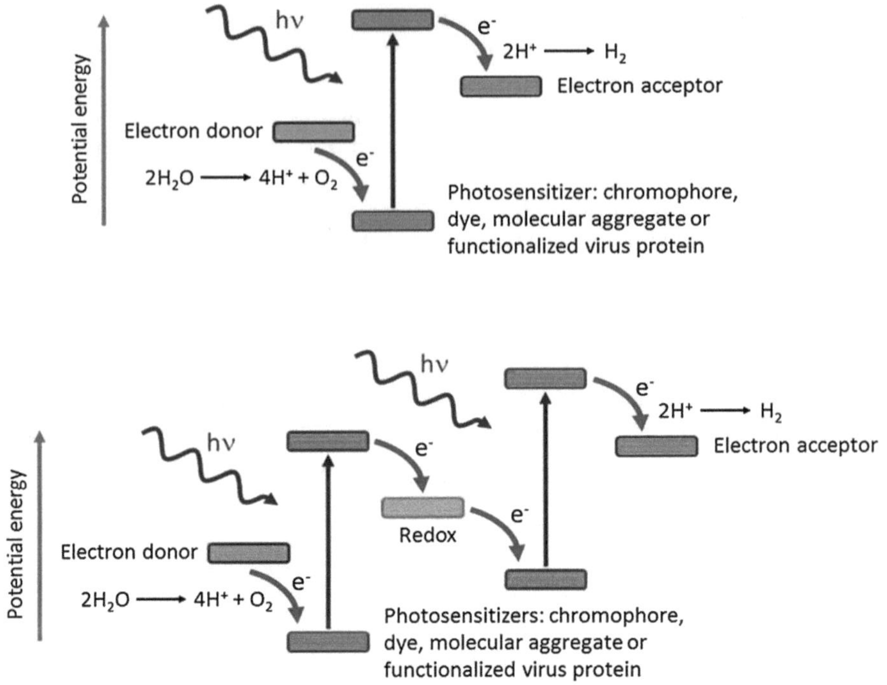

Fig. 10.1 Schematics of the one-step (*top*) and two-step (*bottom*) biomimetic artificial photosynthesis processes. The light-induced charge-separation between donor and acceptor is achieved by replacing natural photosynthetic components with synthesized chromophores, molecular aggregates and functionalized virus proteins

Designing, synthesizing and integrating precisely controlled nanostructures capable of regulating or inhibiting dissipation channels is one of the most challenging fields in contemporary synthetic chemistry.

The simplest way to benefit from the extraordinary capabilities of natural LHCs is to directly incorporate them into a charge collection device that extracts the charge-separated electrons or the photocurrent. These electrons then constitute the detection signal for light sensors or the electrical current in a photovoltaic. This approach has the potential to preserve all the quantum properties of the natural LHCs and was, in fact, first explored by J.C. Bose in the early 1900s. The field of modern "fast" biophotonics, seeking quantitative insights into underlying mechanisms at the molecular level, began in 1932 with publication of the Emerson and Arnold experiments that probed the kinetics and molecular mechanisms of photosynthesis using flashing light [25, 26]. A key finding of the experiments was that 2480 chlorophyll molecules were used for the reduction of one molecule of carbon dioxide per flash of light. This work led directly to the discovery of light-harvesting photosynthetic antennas, which resonates down to our day in the form of ultrafast laser spectroscopy studies on energy migration and conversion in photosynthetic antennas. In this approach, great care needs to be exercised to ensure that the functioning of the LHCs is not adversely affected upon integration into a foreign solid-state environment. Such integration of a natural LHC into a solid-state electron collection device has been accomplished by only a few research groups [27, 28]. They have also succeeded in estimating the internal quantum efficiency of $\sim$12 % with predicted increase to $\sim$20 % by improved integration. The limiting factor in this approach is that very little light is absorbed by the thin monolayer of self-assembled LHCs. Recently, bio-hybrid devices formed by electro-spraying columnar chlorosome structures onto dye-coated surfaces, or by weakly attaching them to electrodes, have yielded encouraging results. Several other studies have focused on the controlled placement and immobilization of natural LHCs onto solid surfaces by complex surface lithography and functionalization techniques [29–31]. A commonly used technique is to form self-assembled monolayers on gold or glass substrates and then to selectively chemically modify them to resist or attract other LHC proteins.

In general, molecular aggregation and Frenkel excitons have been recognized as useful theoretical frameworks for describing, and imitating, many natural photosynthetic LHCs [32–35]. Molecular J-aggregates formed by several types of dye molecules also exhibit similar delocalized collective Frenkel excitons and, hence, are considered as potential candidates for mimicking LHCs [32, 33, 36]. J-aggregates posses unique collective optical properties wherein their optical response is fundamentally different from that of the individual dye molecules. Structural studies carried out on natural LHCs show that their highly efficient antennas potentially consist of cylindrically-shaped aggregates containing several hundred thousand bacteriochlorophyll molecules [5, 7–9, 37]. This has inspired further investigations of light-matter interaction in various artificial π-conjugated molecules and fullerene derivatives that self-assemble to form supramolecular systems acting as either light-harvesting antenna or a reaction centre [38]. Additionally, attempts have been made to produce synthetic molecular aggregates for light-harvesting and energy conversion over a broad range of wavelengths [39–43]. In these systems, the energy conversion generally occurs on ultrafast, femto- to pico-second timescales and was thought to involve incoherent electron transfer processes [32, 33, 35, 44–46]. However, as is discussed in the following, ultrafast studies have highlighted the role of correlated oscillatory motion of electrons and nuclei (vibronic coherence) in charge separation and transfer processes in biomimetic supramolecular reaction centres [20, 21].

An alternative approach to overcoming the design and synthesis challenges is to turn to another natural complex: protein self-assembly [9, 47]. Protein self-assembly is a ubiquitous biological process that occurs at all length scales and with a wide range of structural motifs (sheets, ribbons, helices). Some of the best understood protein self-assembly processes occur in viruses where viral proteins protecting the genetic material self-assemble into a variety of shapes. Self-assembled viral proteins can provide a rigid scaffold for guiding and organizing other molecules. The use of viruses to perform such templated self-assembly has been demonstrated with synthesis of structures for battery technologies [48, 49], targeted drug delivery [50], and for synthesis of artificial light harvesting pigment-protein complexes [51–53]. A favoured virus scaffold used for templating [54] is the tobacco mosaic virus

(TMV). It is particularly appealing for mimicking natural LHCs due to its cylindrical and disc-like geometries [55], closely resembling the natural antenna complex. For light-harvesting applications, the TMV protein is selectively functionalized to facilitate covalent attachment of a wide variety of commercially available chromophores, which in turn mimic donors and acceptors of natural LHCs and produce energetic landscapes for funnelling photon energy. Using such techniques, efficient and robust light-harvesting behaviour has already been demonstrated, with broad spectrum absorption and up to 90 % energy transfer efficiency [51, 52, 56, 57]. Another virus that has been used to template structures with potential for light-harvesting is the M13 bacteriophage [53].

Instead of virus proteins, engineered maquette proteins can also mimick LHCs [58]. These are functionalized self-assembled protein scaffolds designed from scratch to ensure the simplest structure for a given functional task. Like synthetic chemistry, this is a synthetic biology approach that employs iterative and reversible design steps, but the assemblies in general are non-biomimetic. Specific maquette proteins with electron [59] or oxygen transport [60] capabilities have already been demonstrated. A major advantage of protein based light-harvesting structures is their tunability [9]. Chromophore types, densities and orientations can, in principle, be varied and/or adjusted to achieve a desired response for a wide range of applications. Understanding the fundamental trade-offs between different functionalities and developing rational design strategies are essential for fabricating customized devices to collect or sense electromagnetic radiation under diverse conditions. Although the assembled structures of the TMV and several other proteins are well understood, the assembled state after attachment of chromophores is far less known. Detailed structural and dynamical analysis of such assemblies can be performed using electron microscopy techniques (Chap. 3). As with direct molecular aggregates and other approaches, no virus- or maquette-templated assembly of chromophores has, as yet, been integrated into a photovoltaic or a photon-sensing device. However, the incorporation of quantum coherence as an essential characteristic in such natural or engineered protein assemblies may provide additional opportunities for achieving improved light-harvesting or sensing capabilities in an integrated device.

10.3 Quantum Coherence in an Artificial Reaction Centre

Blends of conjugated polymers and fullerene derivatives are prototypical organic photovoltaic materials and potential biomimetic devices for mimicking natural LHCs and their reaction centres. Experiments have shown that the primary charge-generation mechanism in this class of organic photovoltaics involves a light-induced, ultrafast electron transfer from the light-absorbing and electron-donating polymer (equivalent to a chromophore or a pigment) to the fullerene electron acceptor [35, 61, 62]. The currently accepted model for their basic working principle in an organic heterojunction device includes elementary steps like photon absorption by the polymer, generation of Coulomb-bound electron-hole pair (generally Frenkel excitons) in the

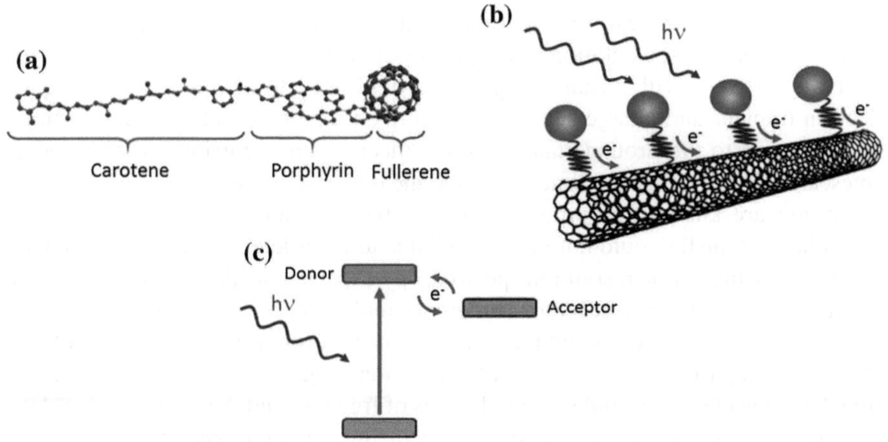

Fig. 10.2 **a** Structure of the carotene-porphyrin-fullerene triad, one of the supramolecules used for fabricating biomimetic photovoltaic devices. **b** Schematic of a biomimetic solar cell comprising chromophores and carbon nanotubes. As in a natural reaction centre, charge separation and transport in such conjugated polymer systems involve quantum dynamics and vibronic coherence between donor and acceptor molecules. **c** Due to coherent coupling, unlike the unidirectional incoherent transfer, the charge oscillates between the donor and the acceptor before being finally transferred to the acceptor

donor phase, exciton diffusion to the donor/acceptor interface, charge separation or dissociation of exciton at the interface and transfer of free charges to the electrodes [61–63]. So far, the photocurrent generation in this class of materials is believed to mainly involve incoherent transfer processes occurring on pico- to femto-second timescale [35, 64–66].

Recently, ultrafast spectroscopic measurements were performed to elucidate the role of quantum dynamics and vibronic coherence in primary charge separation and transfer processes in artificial light-harvesting by polymer-fullerene blends at room temperature [20, 21]. As discussed in Chap. 7, quantum coherence and vibronic modes play a key role in high efficiency quantum mechanical energy and charge transfer processes in photosynthesis [5–10]. Dynamical studies have been carried out on the photoexcited supramolecular carotene-porphyrin-fullerene triad (Fig. 10.2a) and on conjugated polymer-fullerene derivative blended films (poly-3-hexylthiophene-[6,6]-phenyl-C61 butyric acid methyl ester, P3HT:PCBM blended films). A supramolecular carotene-porphyrin-fullerene triad is expected to be more photostable than a simpler porphyrin/fullerene dyad because of the protective role of the carotene in suppressing photobleaching, as in natural reaction centres [20, 21]. The supramolecules consist of a porphyrin/polymer ring as the primary light absorber, a fullerene as an electron acceptor and a carotene group serving as a hole stabilizer. These prototypical materials, depicted in Fig. 10.2a, b, act as biomimetic artificial reaction centres and exhibit typical power conversion efficiencies of ~5–10 % [67–69].

From extensive and careful experimental and theoretical investigations, it seems established that photoexcitation of the porphyrin triggers an ultrafast electron transfer to the fullerene with a charge-separation yield of up to ~95 % [70]. Subsequent electron transfer and charge separation occurs on a picosecond timescale whereas the relaxation to the ground state is a slow mechanism occurring on microsecond timescales [71]. Though the timescales and the major steps involved in photocurrent generation are known, the ultrafast microscopic mechanisms underlying the primary charge-transfer could not be unveiled due to insufficient time resolution. Only very recently high-time resolution spectroscopy experiments demonstrated coherent vibrational motion of the acceptor (fullerene) molecule, on a few tens of femtosecond times, upon impulsive excitation of the donor (porphyrin or polymer) molecule [20, 21]. Comparison with theoretical simulations suggests coherent vibronic coupling between electronic and nuclear degrees of freedom and the resulting correlated oscillatory motion of the wavepackets, as occurring in natural reaction centres, is the primary driving mechanism that triggers charge delocalization and transfer in these prototypical systems.

A consistent and general picture of the elementary photoinduced charge-transfer process in the P3HT:PCBM blend emerges from detailed experimental and theoretical studies [21]. Optical excitation locally creates an electron-hole pair on the polymer moiety. The strong vibronic coupling between electronic and nuclear degrees of freedom promotes a delocalization of the optically excited electronic wave packet across the interface. Both the electronic density and the nuclei display correlated oscillations on the same timescales, essential for ultrafast charge transfer from donor to acceptor. The observation of coherent electron-nuclear motion in a non-covalently bound complex, averaging over a macroscopic ensemble of P3HT:PCBM moieties with variable environment and interfaces, is strong evidence for the dominant role of quantum coherences in the early stages of the charge transfer dynamics in this class of organic photovoltaic materials [20, 21].

An optical excitation locally launches an electron-hole pair wavepacket onto the excited state surface of the absorber. The strong vibronic coupling between electronic and nuclear degrees of freedom induces delocalization of this wavepacket across the interface via coherent nuclear motion of the carbon backbone [20, 21]. Unlike the unidirectional incoherent transfer, the charge oscillates between the absorber and the acceptor with a period of a few tens of femtoseconds (schematically depicted in Fig. 10.2c). The transient charge generation on the acceptor leads to gradual rearrangement of the molecular geometry and a concomitant time-dependent change of the acceptor potential energy surface that helps stabilize the charge accumulation. Theoretical simulations have shown that the vibrational degrees of freedom are essential for charge separation and transfer. Clamping of either donor or acceptor molecules does not result in any charge transfer to the fullerene moiety. Hence, it is the coherent motion of strongly coupled electrons and ions that drives the charge separation, while the surrounding environment apparently plays a less important role [70], in contrast with photosynthesis. As the reader can imagine, the probability for charge transfer is also large whenever the polymer and fullerene molecules are transiently brought into resonance. Taken together, the experimental and theoretical results imply that not only the geometric orientation and electronic structure of a photovoltaic material but also the flexibility of its molecular structure (governed by donor-acceptor linking groups) are important for optimizing the performance of biomimetic photovoltaic devices [20, 21]. It would be interesting to explore optimization of vibronic coupling in developing more efficient and photostable organic photovoltaic materials. Though found to be less affected by the environment, introducing coupling to the surrounding, for example to the polarity of the solvent or the matrix, might provide environment-assisted charge separation and transport, making these devices adaptive to different illumination conditions. It would also be interesting to carry out structural investigations of these supramolecules using time resolved X-ray and/or electron diffraction techniques [7–9], to obtain information about light-induced ultrafast molecular rearrangements of the type that have already been observed in isolated molecules in gas-phase experiments with intense, few-cycle laser pulses [72–74].

10.4 Biomimetic Magnetic Field Sensors

As discussed in Chap. 8, birds navigate by exploiting the quantum-mechanical evolution of photoexcited radical pairs, a process that is dependent on the latitude-dependent inclination of the Earth's magnetic field [8, 75–80]. Although a definitive mechanism of this radical pair production and evolution remains unknown, it can be used to develop artificial devices for magnetometry [81]. Though magnetometry is a well-developed field [82], and many magnetic field sensors are commercially available, there is still a need to develop low-cost, reliable weak magnetic field sensors operating at room temperature. The most sensitive detector of a DC, or low frequency, magnetic field, is a quantum mechanical Josephson effect based SQUID magnetometer which can detect fields of $\sim 10^{-14}$ T; but this needs cooling to cryogenic temperatures and has a slow response time. Nanoscale magnetometers relying on manipulating quantum mechanical spin coherence have recently been constructed using nitrogen-vacancy centres in diamond [83, 84]. Since magnetic field is a vector quantity, a magnetic field sensor should provide information about the overall strength as well as the field direction. However, the latter task—of determining a field direction—is typically more challenging. Though remarkable quantum mechanical SQUID-based magnetic field sensors are already commercially available [82], an ultrafast, nanoscale, low-cost, reliable, room temperature biomimetic magnetic field sensor—or a chemical compass—remains on the wish list.

There are several theoretical and simulation studies demonstrating the radical pair mechanism for detecting weak magnetic fields [81]. However, experimental realization remains to be achieved. Proof-of-principle experiments for demonstrating sensitivity of a photo-activated radical pair reaction in artificial molecules to the magnitude and direction of weak magnetic fields have been reported [85, 86]. The molecule used in these experiments [20] is, helpfully for our discussion of artificial light-harvesting applications, a carotene-porphyrin-fullerene triad (Fig. 10.2a). Figure 10.3 shows a schematic of the set-up used; the orientation of the triad was fixed by immobilizing

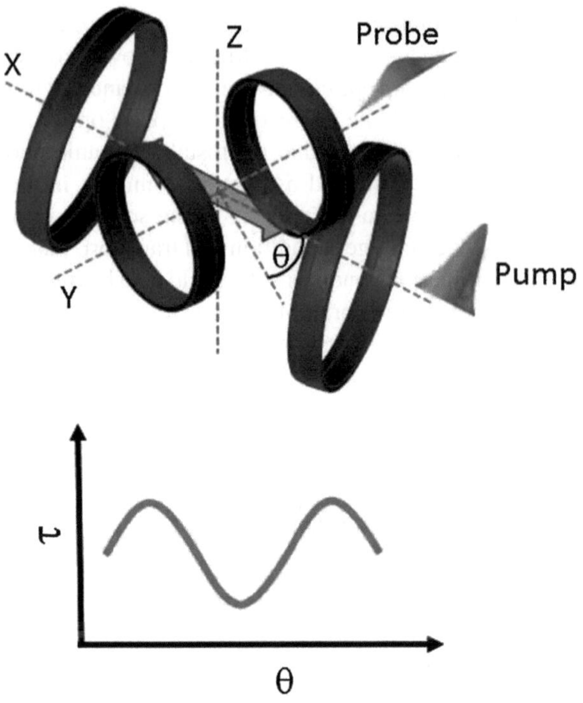

Fig. 10.3 Schematic of the experimental arrangement (*top*) used to measure the anisotropy of the magnetic field dependence of carotene-porphyrin-fullerene triad dynamics (see text). An anisotropic magnetic field was generated in the x–y plane by two pairs of Helmholtz coils. The triad, represented by the *red arrow*, was immobilized in a liquid crystal matrix with its axis parallel to the x-direction. The pump and probe pulses were also polarized in the horizontal (x–y) plane. The relaxation time of the photochemically formed radical pair, measured using pump-probe spectroscopy, exhibits a periodic variation as a function of angle θ (*bottom*), reflecting the magnetic field anisotropy. Such field dependent variation in the photochemically generated radical pair lifetime can be exploited as a biomimetic chemical compass, sensitive to the magnitude and direction of the magnetic field

it in the nematic phase of a liquid crystal [85]. Upon green-light irradiation, the triad generates a delocalized exciton-hole pair (Frenkel exciton) extending over the entire extent of the supramolecule [32–35]. The magnetic field dependence of the sub-microsecond recombination rate of this radical pair state, recorded via transient absorption measurements at 113 K, demonstrated a high sensitivity towards magnetic field strength and direction, being able to detect field magnitude variations down to ~50 μT, and directional variations down to mT [85]. A greater sensitivity, faster response times and higher operation temperatures are expected with better optimization of the molecular composition of the chemical compass. For comparison, the magnitude of the Earth's magnetic field varies between ~20 and ~50 μT. The main question to be addressed for the development of such biomimetic magnetic sensors is the optimization of the molecular structure for the generation of the photoexcited radical pair. The photoreceptor that hosts the radical pair mechanism in birds is the cryptochrome flavoprotein [8, 75–80]. It would be interesting to try this structure, or its variants, for radical pair production. The desirable factors for this purpose include a molecular structure supporting sufficiently small separation between the photoexcited radical pair, optimum radical pair relaxation time and its magnetic environment [79, 86–88]. The radical pair lifetime should be long enough for the weak magnetic field to affect coherent recombination dynamics, but should be short compared to other incoherent spin relaxation dynamics which are insensitive to magnetic field effects. Some designs to introduce magnetic nanostructures which not only immobilize the chemical compass but also enhance the field sensitivity and optimize the recombination rate have already been proposed [89]. Fabrication of a prototypical quantum mechanical magnetic compass is still an open field and highlights the need for a systematic study of molecular structures, efficient radical pair production and its dynamics. Real-time visualization of structural reorientation of molecular structure caused by the magnetic field will be important in optimizing device architecture.

10.5 Biomimetic Bacteriorhodopsin Devices

As noted in earlier chapters, biological systems evolve a variety of functionalities that provide comparative advantages. In Chaps. 7 and 8, we have come across a relatively rare class of photochromic proteins designed to convert light stimuli into a physiological signal. One such light-conversion mechanism is that of isomerization (*trans* to *cis* or *cis* to *trans*) of a carbon-carbon double bond in an extended conjugated chromophore system, that ultimately leads to a biological response [90, 91]. As already noted, these photochromic proteins play a key role in photosynthesis, magnetoreception, and in visual perception. Due to their importance, numerous studies have been undertaken to elucidate their structure, optical response, and the mechanisms of how light energy is converted into chemical energy or into physiological signals [8, 90]. In several cases, the microscopic mechanisms have been found to occur on ultrafast timescale and involve quantum mechanical processes [90–93]. One of the best known biological photoreceptors is the rhodopsin molecule with its retinal chromophore (Chap. 8) that is responsible for the triggering vision. Microbial rhodopsins present in various unicellular organisms have similar functionality, but they carry out two distinctly different functions: light-driven ion transport and photosensory signalling that controls swimming [90]. A rather simple representative of the class of microbial rhodopsins, bacteriorhodopsin (bR), was discovered about four decades ago [94]. It is a highly stable light-activated proton pump produced by halobacteria and is the key protein for photosynthetic capabilities in halobacteria. bR is one of the most studied proteins for use in bioelectronic and biomimetic devices [95–97]. Soon after its discovery, proposals for biomimetic photochromic applications of bR in all-optical switching, three-dimensional optical memories, real-time holographic processors and artificial retinas, were explored [95, 96, 98–103]. Though most ultrafast studies have been performed on bR, the results are also applicable to other rhodopsins because of structural and functional similarities.

An important question for biomimetic applications is: what makes bR so attractive over other proteins and conventional inorganic or organic photochromic materials? The key to the answer lies in the physical mechanisms of bR, which are understood on a molecular level, and are used as genetic tools to redesign the protein [95, 96]. Several investigations have shown that even naturally occurring bR has quite attractive properties: unlike in photosynthesis, only a single bR molecule is needed to convert light energy into chemical energy [104, 105]. For photochromic applications, its photosensitivity and photostability are far better than synthetic materials. It occurs as a two dimensional crystal structure, which is extremely stable against chemical and thermal degradation; it can also be modified to a large extent, thereby serving as a platform for a whole new class of materials [95, 96]. These characteristics make bR not only an attractive candidate for biomimetic applications but also for investigations of other biomaterials.

The bR protein, comprising seven transmembrane alpha helices, is arranged in a two-dimensional hexagonal lattice of trimers in a lipid bilayer that is often referred to as the purple membrane [106–109]. The inherent stability of the semi-crystalline matrix allows the protein to survive and remain functionally active in the most extreme environments. When environmental oxygen becomes scarce and drops below a level necessary to sustain oxidative phosphorylation, the bacteria express bR to generate energy via photosynthesis [109]. The absorption of a photon by the retinal chromophore of bR is one of the most efficient and stable photochemical reactions found in nature [110, 111], with a quantum efficiency of 65 % for the primary step [112]. After decades of research, the photocyclic mechanism of proton pumping of bR is well characterized and several models have been proposed to explain its complex photocycle, made up of a series of transient photochemical and conformational intermediates [113–116]. The primary light-absorbing moiety of bR is an all-*trans* retinal chromophore that is bound to the protein via a protonated base linkage. Absorption of light by the retinal chromophore causes it to isomerize (all-*trans* to 13-*cis* for bR and 11-*cis* to all-*trans* for rhodopsin). Only the first step of the entire bR photocycle is light driven; all the other steps are dark reactions [90]. Ultrafast studies on bR and rhodopsin, with high time resolution, have identified several spectroscopic intermediates preceding the first stable photoproduct [92, 93]: isomerization in bR is, indeed, very fast (200–500 fs) and can be a potential candidate for applications in ultrafast biophotonics [95, 96]. The basic molecular functions of the bR photocycle and their corresponding physical effects are schematically shown in Fig. 10.4. The light-driven isomerization is followed immediately by a charge separation and proton transport step accounting for the photoelectric properties of bR. The charge transport through the molecule requires deprotonation and re-protonation of the base linkage in the photochromic group, causing a shift in the absorption maximum of more than 150 nm. Most of the intermediate states of the bR photocycle are characterized by unique optical and proton transport properties making it promising for a wide range of biomimetic applications like photochromic optical recording, artificial photosynthesis and seawater desalination [95, 96].

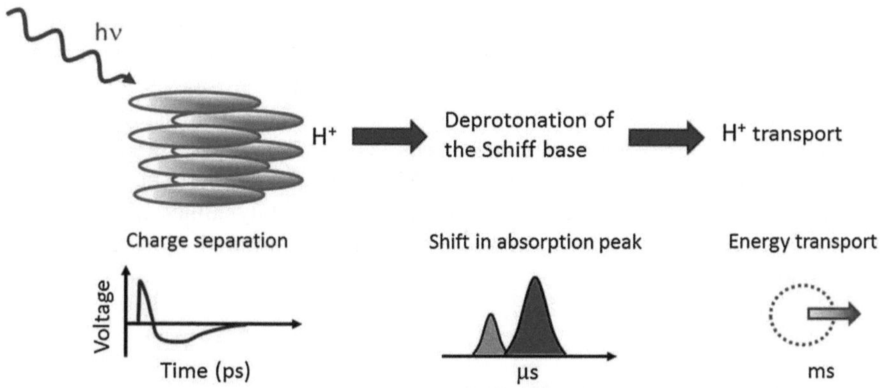

Fig. 10.4 Basic molecular functions of the bacteriorhodopsin (bR) photocycle. Only the first step of the entire cycle, involving all-*trans* to 13-*cis* isomerization, is light-driven and occurs on femtosecond timescales; others are "dark" reactions that occur on slower timescales. The isomerization is accompanied by proton transport and a charge separation step on the picosecond timescale. After about 50 μs the deprotonation of the Schiff base leads to the main photochromic shift during the photocycle. The proton as well as energy transport are completed on the millisecond timescale. Each step of the photocycle can be used to develop biomimetic devices for chemical information processing at the molecular or supramolecular level

bR has emerged as one of the most rigorously investigated biomaterial for use in electronic applications, culminating in the development of numerous prototype devices that include Fourier transform holographic associative processors, three-dimensional optical memories, optical bistability and switching, ultrafast light detection, biosensors and photovoltaic cells [95–97, 117]. Its domain of applications include encoding, manipulation and retrieval of the chemical information at the molecular or supramolecular level. Recent advances in genetic engineering and directed evolution have further helped in optimizing these protein-based devices and enhancing the potential of bR-based technology. An interesting application of bR in biomedical devices is that of protein-based retinal prosthesis for patients suffering from retinal degenerative diseases such as age-related macular degeneration and retinitis pigmentosa. The proposed retinal implant mimics the light-absorbing capabilities of the homologous native visual pigments simulating the neural network of the damaged retina [118]. Apart from device applications, bR is also used as a model system for studying and optimization of optical, electronic, thermal and electronic properties of other engineered biomaterials [95].

Though naturally-occurring bR has many interesting properties, it cannot be regarded as an *optimized* molecule for use in artificial devices [95, 96]. In its native state, this protein does not function efficiently for a branched and/or partial photocycle. Recent advancements in genetic engineering, particularly directed evolution, allows for stepwise manipulation of the properties of natural proteins, promoting their use in nanotechnology [119, 120]. Similar optimization of bR for a synthetic environment requires enhancing properties of the protein, which are either irrelevant or non-essential for biological function, and suppressing those which are essential in the natural environment. Although researchers are actively studying the structural properties and intramolecular interactions relevant to bR, many of these mechanisms still remain unknown [95, 96]. Hence, despite having access to advanced nanotechnology and rigorous theoretical models, it is very challenging to predict an ideal protein structure for a specific application. More extensive experimental investigations to determine protein structure as well as its ultrafast dynamics and theoretical modeling are needed in order to make protein based biomimetic devices commercially viable.

10.6 Biomolecular Electronics

Molecular electronics is at the heart of nanoscience [121–124]. Molecules are intrinsically quantum mechanical and, consequently, molecular devices exhibit characteristics that cannot be achieved by conventional classical devices. Organic molecules have now been successfully incorporated in electronics in the form of organic light-emitting diodes and in molecule-based photovoltaics, but the practical use of a single-molecule as a "molecular transistor" remains elusive. Though DNA-based field-effect transistors have been proposed and demonstrated [125–127], a biomimetic single molecule transistor remains a challenge; most attempts have aimed at replicating the functionality of a semiconductor transistor with a single organic molecule [128]. Another major challenge is to interface molecules with metallic leads. Carbon-based electrodes seem to be a promising approach, because in this case the molecules and electrodes are of the same material and, hence, carbon-carbon bonds will considerably reduce electronic mismatch at the junction interface. To this end, cage macromolecular structures or supramolecular aggregates are potential candidates. Driven by recent breakthroughs in both experiment and theory, single-molecule junctions are emerging as novel ensemble platforms for understanding electronic structure and function at metal-molecular interfaces. New and exciting phenomena—such as thermoelectric effects, mechanical switching, quantum interference and spin filtering—have been reported at the single-molecule level. Biological macromolecules have, over millions of years, evolved into structures that are primed for both specific surface recognition and directional electron tunneling. As we have seen, the redox-active centres of metalloproteins like myoglobin play a central role in respiration, photosynthesis, magneto-reception, and vision [129, 130]. Together with recent advances in molecular manipulation and lithographic fabrication, understanding of the protein functionalities has led to the point where bioelectronic devices can be designed and fabricated with good levels of reproducibility [131].

The efficiency with which some of the metalloproteins mediate ultrafast electron tunnelling is a key to sustaining processes like respiration [129, 130]. Electron-transfer reactions between donor and acceptor structures (typically both redox-active sites) separated by distances well beyond van der Waals interactions are common in nature [59, 132]. Here, the transfer process, as discussed in Chap. 9, is usually

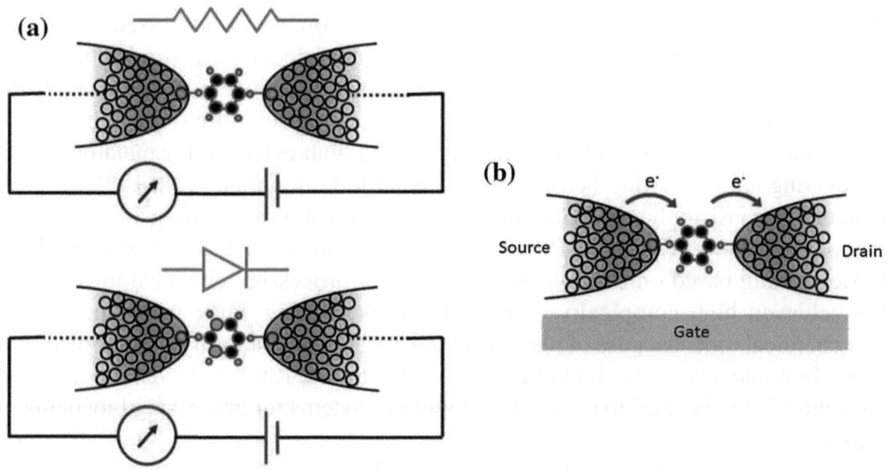

Fig. 10.5 **a** Schematic of a single-biomolecular junction diode. The symmetric biomolecule (*top*) allows for two-way current, acting as a resistor, whereas the asymmetrical biomolecule (*bottom*) permits current in one direction only, thereby acting as a single-molecule diode. The device can be switched between the two states by applying an ultrashort light pulse of appropriate wavelength. **b** Schematic of a single-biomolecular tunneling transistor. Electron tunneling from source to drain via a biomolecule, like a metalloprotein, is controlled by an ultrashort optical or electronic gate pulse

non-adiabatic with the barrier potential being much larger than thermal energy at room temperature. The weak electronic coupling is mediated by the potential energy landscape of the intervening medium. The coupling reduces exponentially with distance between the acceptor and donor molecules. Metalloproteins offer several specific advantages over other organic molecules in fabrication of tunnel junctions. In particular, these proteins are specifically designed to facilitate efficient electron tunneling and their primary role in the natural environment is in electron-transfer [133]. Thus, their intrinsic structural properties make them interesting potential candidates for pursuing biomolecular electronics and biomolecular computing [134–136]. Schematics of the typical single-biomolecular junction device [121, 122] are shown in Fig. 10.5. Some of the metalloproteins have redox-active metal centres which provide a mechanism for the molecular electronic structure to be switched (changing the oxidation state of the transition metal centre) by an external optical or electronic "gate" [137]. In many cases, the metal component of these metalloproteins can be substituted with negligible change in geometric structure, facilitating examination of the precise role of the metal centre. Finally, like many biomolecules, these proteins exhibit interfacial recognition characteristics that may be optimized through genetic engineering for developing highly efficient sensors [122, 131, 137].

Though many synthetic molecules have been proposed for molecular electronics [122, 137], developments in this field have met with limited success due to the inherent complexity of synthesizing precise molecular networks. However, such synthetic chemistry related problems may be more easily addressed by utilization of biomolecular systems. Another field of biomolecular electronics has, until very recently, been largely dominated by harnessing the molecular transport properties of the DNA molecule [131]. Due to its complex structure, reliable transport measurements have proved to be challenging although evidence for metallic, semi-conducting and insulating behaviour has begun to be reported [138–142]. A DNA field effect transistor has been demonstrated for applications in biosensing and information processing [125–127]. Recent developments in the emerging research field of biocomputing based on biomolecular systems for processing chemical information has achieved high complexity of information processing while using much simpler chemical tools because of natural specificity and compatibility of biomolecules. Other biomolecular tools, including proteins/enzymes, RNA and even whole cells, are beginning to be used to assemble computing systems for processing biochemical information [136].

Interest in biomolecular electronics highlights the need for better understanding of conduction mechanisms in organic materials in synthetic environments [122, 137]. Examples outlined in this section are the results of recent attempts to use single biological macromolecules and may, ultimately, yield a biomolecular device. Future work must, necessarily, include investigations of a variety of lithographically fabricated device configurations, their response and dynamics [131]. The prospects of using the evolved characteristics of biological macromolecules are exciting and promising for developing biomimetic single-molecule devices. In order to successfully demonstrate a biomolecular tunnel junction, it is imperative to understand the dominant mechanisms of charge transport and the environmental impacts. Single molecule spectroscopy and imaging techniques discussed in Chap. 4 will play an important role in achieving these goals. Other factors like ionic, electrolytic and hydrogen-bonding in biomolecular tunnel junctions might also be interesting to pursue.

References

1. A. von Gleich, C. Pade, U. Petschow, E. Pissarskoi, *Potentials and Trends in Biomimetics* (Springer, Berlin, 2010)
2. V. Giovannetti, S. Lloyd, L. Maccone, Nat. Photon. **5**, 222 (2011)
3. M.A. Nielsen, I.L. Chuang, *Quantum Computation and Quantum Information* (Cambridge University Press, Cambridge, 2011)
4. N. Gisin, R. Thew, Nat. Photon. **1**, 165 (2007)
5. G.S. Engel, T.R. Calhoun, E.L. Read, T.-K. Ahn, T. Mančal et al., Nature **446**, 782 (2007)
6. P.C.W. Davies, *Quantum Aspects of Life* (Imperial College Press, London, 2008)
7. M. Sarovar, A. Ishizaki, G.R. Fleming, K.B. Whaley, Nat. Phys. **6**, 462 (2010)
8. N. Lambert, Y.-N. Chen, Y.-C. Cheng, C.-M. Li, G.-Y. Chen, F. Nori, Nat. Phys. **9**, 10 (2013)
9. M. Mohseni, Y. Omar, G.S. Engel, M.B. Plenio (eds.), *Quantum Effects in Biology* (Cambridge University Press, Cambridge, 2014)
10. E. Romero, R. Augulis, V.I. Novoderezhkin, M. Ferretti, J. Thieme et al., Nat. Phys. **10**, 676 (2014)

11. G.D. Scholes, Nat. Phys. **7**, 448 (2011)
12. B. Li, L. Wang, B. Kang, P. Wang, Y. Qiu, Sol. Energy Mat. Sol. C. **90**, 549 (2006)
13. M. Grätzel, Nature **414**, 338 (2001)
14. A. Ishizaki, G.R. Fleming, Proc. Natl. Acad. Sci. USA **106**, 17255 (2009)
15. D. Gust, T.A. Moore, A.L. Moore, Acc. Chem. Res. **42**, 1890 (2009)
16. G.D. Scholes, G.R. Fleming, A. Olya-Castro, R. van Grondelle, Nat. Chem. **3**, 763 (2011)
17. R.E. Blankenship, D.M. Tiede, J. Barber, G.W. Brudvig, G. Fleming et al., Science **332**, 805 (2011)
18. Y. Tachibana, L. Vayssieres, J.R. Durrant, Nat. Photon. **6**, 511 (2012)
19. J.H. Kim, D.H. Nam, C.B. Park, Curr. Opin. Chem. Biol. **28**, 1 (2014)
20. C.A. Rozzi, S.M. Falke, N. Spallanzani, A. Rubio, E. Molinari et al., Nat. Commun. **4**, 1602 (2013)
21. S.M. Falke, C.A. Rozzi, D. Brida, M. Maiuri, M. Amato et al., Science **344**, 1001 (2014)
22. A.K. Manske, J. Glaeser, M.M.M. Kuypers, J. Overmann, Appl. Environ. Microbiol. **71**, 8049 (2005)
23. E.H. Murchie, K.K. Niyogi, J. Plant Physiol. **155**, 86 (2011)
24. K. Norrman, N.B. Larsen, F.C. Krebs, Sol. Energ. Mat. Sol. Cells **90**, 2793 (2006)
25. R. Emerson, W. Arnold, J. Gen. Physiol. **15**, 391 (1932)
26. R. Emerson, W. Arnold, J. Gen. Physiol. **16**, 191 (1932)
27. R. Das, P.J. Kiley, M. Segal, J. Norville, A.A. Yu et al., Nano Lett. **4**, 1079 (2004)
28. S.A. Trammell, L. Wang, J.M. Zullo, R. Shashidhar, N. Lebedev, Biosens. Bioelectron. **19**, 1629 (2004)
29. M. Ciobanu, H.A. Kincaid, G.K. Jennings, D.E. Cliffel, Langmuir **21**, 692 (2005)
30. N.P. Reynolds, S. Janusz, M. Escalante-Marun, J. Timney, R.E. Ducker et al., J. Am. Chem. Soc. **129**, 14625 (2007)
31. M. Escalante, P. Maury, C.M. Bruinink, K. van der Werf, J.D. Olsen et al., Nanotechnology **19**, 025101 (2008)
32. T. Kobayashi, *J-Aggregates* (World Scientific, Singapore, 1996)
33. H. van Amerongen, L. Valkunas, R. van Grondelle, *Photosynthetic Excitons* (World Scientific, Singapore, 2000)
34. R.E. Blankenship, *Molecular Mechanisms of Photosynthesis*, 2nd edn. (Wiley-Blackwell, Hoboken, 2014)
35. G. Lanzani, *The Photophysics Behind Photovoltaics and Photonics* (Wiley-VCH, Weinheim, 2012)
36. G.D. Scholes, Chem. Phys. **275**, 373 (2002)
37. G.D. Scholes, Nat. Phys. **6**, 402 (2010)
38. F. Würthner, T.E. Kaiser, C.R. Saha-Moller, Angew. Chem. Int. Ed. **50**, 3376 (2011)
39. B.J. van Rossum, D.B. Steensgaard, F.M. Mulder, G.-J. Boender, K. Schaffner et al., Biochemistry **40**, 1587 (2001)
40. S.M. Vlaming, R. Augulis, M.C.A. Stuart, J. Knoester, P.H.M. van Loosdrecht, J. Phys. Chem. **113**, 2273 (2009)
41. H. Shao, M. Gao, S.H. Kim, C.P. Jaroniec, J.R. Parquette, Chem. Eur. J. **17**, 12882 (2011)
42. T. Aida, E.W. Meijer, S.I. Stupp, Science **335**, 813 (2012)
43. M.R. Wasielewski, Acc. Chem. Res. **42**, 1910 (2009)
44. D.M. Eisele, J. Knoester, S. Kirstein, J.P. Rabe, D.A. Vanden Bout, Nat. Nanotech. **4**, 658 (2009)
45. D.M. Eisele, C.W. Cone, E.A. Bloemsma, S.M. Vlaming, C.G.F. van der Kwaak et al., Nat. Chem. **4**, 655 (2012)
46. V.I. Prokhorenko, A.R. Holzwarth, M.G. Müller, K. Schaffner, T. Miyatake, H. Tamiaki, J. Phys. Chem. B **106**, 5761 (2002)
47. G.M. Whitesides, J.P. Mathias, C.T. Seto, Science **254**, 1312 (1991)
48. K.T. Nam, D.-W. Kim, P.J. Yoo, C.-Y. Chiang, N. Meethong et al., Science **312**, 885 (2006)
49. E. Royston, A. Ghosh, P. Kofinas, M.T. Harris, J.N. Culver, Langmuir **24**, 906 (2008)
50. W. Wu, S.C. Hsiao, Z.M. Carrico, M.B. Francis, Angew. Chem. Int. Ed. **48**, 9493 (2009)

51. M. Endo, M. Fujitsuka, T. Majima, Chem. Eur. J. **13**, 8660 (2007)
52. R.A. Miller, A.D. Presley, M.B. Francis, J. Am. Chem. Soc. **129**, 3104 (2007)
53. Y.S. Nam, T. Shin, H. Park, A.P. Magyar, K. Choi et al., J. Am. Chem. Soc. **132**, 1462 (2010)
54. H. Fraenkel-Conrat, R.C. Williams, Proc. Natl. Acad. Sci. USA **41**, 690 (1955)
55. A. Klug, Philos. T. R. Soc. Lond. B **354**, 531 (1999)
56. R.A. Miller, N. Stephanopoulos, J.M. McFarland, A.S. Rosko, P.L. Geissler, M.B. Francis, J. Am. Chem. Soc. **132**, 6068 (2010)
57. M. Sarovar, K.B. Whaley, New J. Phys. **15**, 013030 (2013)
58. P.L. Dutton, C.C. Moser, Faraday Discuss. **148**, 443 (2011)
59. C.C. Page, C.C. Moser, X. Chen, P.L. Dutton, Nature **402**, 47 (1999)
60. R.L. Koder, J.L.R. Anderson, L.A. Solomon, K.S. Reddy, C.C. Moser, P.L. Dutton, Nature **458**, 305 (2009)
61. G. Yu, J. Gao, J.C. Hummelen, F. Wudl, A.J. Heeger, Science **270**, 1789 (1995)
62. C.J. Brabec, S. Gowrisanker, J.J.M. Halls, D. Laird, S. Jia, S.P. Williams, Adv. Mater. **22**, 3839 (2010)
63. I.-W. Hwang, D. Moses, A.J. Heeger, J. Phys. Chem. C **112**, 4350 (2008)
64. C.J. Brabec, G. Zerza, G. Cerullo, S. De Silvestri, S. Luzzati et al., Chem. Phys. Lett. **340**, 232 (2001)
65. J. Guo, H. Ohkita, H. Benten, S. Ito, J. Am. Chem. Soc. **132**, 6154 (2010)
66. G. Grancini, M. Maiuri, D. Fazzi, A. Petrozza, H.-J. Egelhaaf et al., Nat. Mater. **12**, 29 (2013)
67. W. Ma, C. Yang, X. Gong, K. Lee, A.J. Heeger, Adv. Funct. Mater. **15**, 1617 (2005)
68. B.A. Baker, J.H. Choi, Nano **4**, 189 (2009)
69. M.T. Dang, L. Hirsch, G. Wantz, Adv. Mater. **23**, 3597 (2011)
70. G. Kodis, P.A. Liddell, A.L. Moore, T.A. Moore, D. Gust, J. Phys. Org. Chem. **17**, 724 (2004)
71. D. Gust, T.A. Moore, A.L. Moore, Acc. Chem. Res. **34**, 40 (2001)
72. M. Krishnamurthy, F.A. Rajgara, D. Mathur, J. Chem. Phys. **121**, 9765 (2004)
73. M. Garg, A.K. Tiwari, D. Mathur, J. Chem. Phys. **136**, 024320 (2012)
74. F.A. Rajgara, D. Mathur, A.K. Dharmadhikari, C.P. Safvan, J. Chem. Phys. **130**, 231104 (2009)
75. C.T. Rodgers, P.J. Hore, Proc. Natl. Acad. Sci. USA **106**, 353 (2009)
76. K. Schulten, C.E. Swenberg, A. Weller, Z. Phys, Chem. **111**, 1 (1978)
77. T. Ritz, P. Thalau, J.B. Phillips, R. Wiltschko, W. Wiltschko, Nature **429**, 177 (2004)
78. T. Ritz, M. Ahmad, H. Mouritsen, R. Wiltschko, W. Wiltschko, J.R. Soc. Interface **7**, S135 (2010)
79. E.M. Gauger, E. Rieper, J.J.L. Morton, S.C. Benjamin, V. Vedral, Phys. Rev. Lett. **106**, 040503 (2010)
80. K. Maeda, A.J. Robinson, K.B. Henbest, H.J. Hogben, T. Biskup et al., Proc. Natl. Acad. Sci. USA **109**, 4774 (2012)
81. J.C. Weaver, T.E. Vaughan, R.D. Astumian, Nature **405**, 707 (2000)
82. A. Edelstein, J. Phys. Condens. Matter **19**, 165217 (2007)
83. D. Le Sage, K. Arai, D.R. Glenn, S.J. DeVience, L.M. Pham et al., Nature **496**, 486 (2013)
84. G. Balasubramanian, I.Y. Chan, R. Kolesov, M. Al-Hmoud, J. Tisler et al., Nature **455**, 648 (2008)
85. K. Maeda, K.B. Henbest, F. Cintolesi, I. Kuprov, C.T. Rodgers et al., Nature **453**, 387 (2008)
86. C.T. Rodgers, S.A. Norman, K.B. Henbest, C.R. Timmel, P.J. Hore, J. Am. Chem. Soc. **129**, 6746 (2007)
87. C.R. Timmel, U. Till, B. Brocklehurst, K.A. Mclauchlan, P.J. Hore, Mol. Phys. **95**, 71 (1998)
88. O. Effimova, P.J. Hore, Biophys. J. **94**, 1565 (2008)
89. J.-M. Cai, Phys. Rev. Lett. **106**, 100501 (2011)
90. V. Sundström, Annu. Rev. Phys. Chem. **59**, 53 (2008)
91. M.A. van der Horst, K.J. Hellingwerf, Acc. Chem. Res. **37**, 13 (2004)
92. R.A. Mathies, C.H.B. Cruz, W.T. Pollard, C.V. Shank, Science **240**, 777 (1988)
93. R.A. Mathies, S.W. Lin, J.B. Ames, W.T. Pollard, Rev. Biophys. Biophys. Chem. **20**, 491 (1991)

94. D. Oesterhelt, W. Stoeckenius, Nat. New Biol. **233**, 149 (1971)
95. N. Hampp, Chem. Rev. **100**, 1755 (2000)
96. N.L. Wagner, J.A. Greco, M.J. Ranaghan, R.R. Birge, J.R. Soc. Interface **10**, 20130197 (2013)
97. R.R. Birge, Annu. Rev. Phys. Chem. **41**, 683 (1990)
98. Q. Li, J.A. Stuart, R.R. Birge, J. Xu, A. Stickrath, P. Bhattacharya, Biosens. Bioelectron. **15**, 869 (2004)
99. R.R. Birge, N.B. Gillespie, E.W. Izaguirre, A. Kusnetzow, A.F. Lawrence et al., J. Phys. Chem. B **103**, 10746 (1999)
100. N. Hampp, Appl. Microbiol. Biotechnol. **53**, 633 (2000)
101. Z. Chen, D. Govender, R. Gross, R. Birge, BioSys. **35**, 145 (1995)
102. Z. Chen, R.R. Birge, Trends Biotechnol. **11**, 292 (1993)
103. S. Roy, P. Sharma, A.K. Dharmadhikari, D. Mathur, Opt. Commun. **237**, 251 (2004)
104. U. Haupts, J. Tittor, D. Oesterhelt, Annu. Rev. Biophys. Biomol. Struct. **28**, 367 (1999)
105. J.B.C. Findlay, D.J.C. Pappin, Biochem. J. **238**, 625 (1986)
106. A.E. Blaurock, W. Stoeckenius, Nat. New Biol. **233**, 152 (1971)
107. A.E. Blaurock, J. Mol. Biol. **93**, 139 (1975)
108. R. Henderson, J. Mol. Biol. **93**, 123 (1975)
109. J.K. Lanyi, Annu. Rev. Physiol. **66**, 665 (2004)
110. R.R. Birge, Annu. Rev. Biophys. Bioeng. **10**, 315 (1981)
111. R.R. Birge, Biochim. Biophys. Acta **1016**, 293 (1990)
112. R.R. Birge, T.M. Cooper, A.F. Lawrence, M.B. Masthay, C. Vasilakis et al., J. Am. Chem. Soc. **111**, 4063 (1989)
113. U. Haupts, J. Tittor, E. Bamberg, D. Oesterhelt, Biochemistry **36**, 2 (1997)
114. J.K. Lanyi, J. Struct. Biol. **124**, 164 (1998)
115. L. Brown, A. Dioumaev, R. Needleman, J. Lanyi, Biochemistry **37**, 3982 (1998)
116. H. Kandori, Biochim. Biophys. Acta **1658**, 72 (2004)
117. N. Hampp, A. Silber, Pure Appl. Chem. **68**, 1361 (1996)
118. N.L. Wagner, J.A. Greco, R.R. Birge, *Bionanotechnology: Biological Self-assembly and Its Applications* (Caister Academic Press, Norfolk, 2013)
119. F. Arnold, Acc. Chem. Res. **31**, 125 (1998)
120. J.R. Hillebrecht, K.J. Wise, R.R. Birge, Method Enzymol. **388**, 333 (2004)
121. A. Aviram, M.A. Ratner, Chem. Phys. Lett. **29**, 277 (1974)
122. J.C. Cuevas, E. Scheer, *Molecular Electronics: An Introduction to Theory and Experiment* (World Scientific, Singapore, 2010)
123. R.L. McCreery, A.J. Bergren, Adv. Mater. **21**, 4303 (2009)
124. Y. Xue, S. Datta, M.A. Ratner, Chem. Phys. **281**, 151 (2002)
125. G. Maruccio, P. Visconti, V. Arima, S. D'Amico, A. Biasco et al., Nano Lett. **3**, 479 (2003)
126. H. Simchi, M. Esmaeilzadeh, H. Mazidabadi, J. Appl. Phys. **113**, 074701 (2013)
127. S.J. Tans, A.R.M. Verschueren, C. Dekker, Nature **393**, 49 (1998)
128. E. Lörtscher, Nat. Nanotech. **8**, 381 (2013)
129. J.W. Petrich, C. Poyart, J.L. Martin, Biochem. **27**, 4049 (1988)
130. M. Levantino, G. Schirò, H.T. Lemke, G. Cottone, J.M. Glownia et al., Nat. Commun. **6**, 6772 (2015)
131. J.J. Davis, Phil. Trans. R. Soc. A **361**, 2807 (2003)
132. J.J. Regan, J.N. Onuchic, Adv. Chem. Phys. **107**, 497 (1999)
133. J.J. Davis, H.A.O. Hill, A. Kurz, C. Jacob, W. Maret, B.L. Vallee, Phys. Chem. Comm. **1**, 12 (1998)
134. J.M. Beebe, V.B. Engelkes, L.L. Miller, C.D. Frisbie, J. Am. Chem. Soc. **124**, 11268 (2002)
135. X.D. Cui, X. Zarate, J. Tomfohr, O.F. Sankey, A. Primak et al., Science **294**, 571 (2001)
136. S. Sanvito, Chem. Soc. Rev. **40**, 3336 (2011)
137. S.V. Aradhya, L. Venkataraman, Nat. Nanotech. **8**, 399 (2013)
138. E. Meggers, M.E. Michel-Beyerle, B. Geise, J. Am. Chem. Soc. **120**, 12950 (1998)
139. Y. Okahata, T. Kobayashi, K. Tanaka, M.J. Shimomura, J. Am. Chem. Soc. **120**, 6165 (1998)
140. H.W. Fink, C. Schonenberger, Nature **398**, 407 (1999)
141. D. Porath, A. Bezryadin, S. de Vries, C. Dekker, Nature **403**, 635 (2000)
142. A.J. Storm, J. van Noort, S. de Vries, C. Dekker, Appl. Phys. Lett. **79**, 3881 (2001)

Chapter 11
Future Opportunities

Abstract In this chapter we present an overview of the prospects of forthcoming experimental and theoretical advances in areas like the use of femtosecond X-ray pulses from a free electron laser for ultrashort imaging of biological materials without causing sample damage, X-ray scattering interferometry, application of terahertz radiation, frequency combs for broadband high-precision spectroscopy, coherently controlling biochemical reactions and explorations of biological entities beyond the quantum limit.

11.1 Outlook

Experimental as well as theoretical methods developed in fields of research as diverse as femtochemistry, semiconductor physics, statistical mechanics and quantum information processing, among others, are likely to open up new directions of research and development in ultrafast biophotonics. For instance, single-molecule studies (Chap. 4) have already started providing remarkable new insights into functions of biological systems [1, 2]. These phenomena were previously hidden in the ensemble average common to almost all earlier studies. Investigations of quantum-dynamical phenomena at the single-molecule level are very challenging because of the ultrashort timescales involved. However, recent observations of vibrational coherence in individual molecules at ambient temperatures by means of a phase-locked spontaneous fluorescence technique auger well for future developments [2]. Application of this technique to detect electronic coherence in photosynthetic complexes (Chaps. 4 and 7) and photoreceptors (Chap. 8) may provide important new insights into quantum energy flow in biology. In the following we present an overview of some other topics which, in the authors' opinions, are likely to be vigorously pursued in the coming years. The prognosis is extremely good for qualitative leaps to be attained vis-a-vis development of new insights into biophotonics.

P. Vasa and D. Mathur, *Ultrafast Biophotonics*, Biological and Medical Physics, Biomedical Engineering, DOI 10.1007/978-3-319-39614-9_11

11.2 Femtosecond X-Ray Imaging

The advent of femtosecond-long hard X-rays from free electron laser (FEL) facilities, such as the Linac Coherent Light Source (LCLS) in Stanford, FLASH (Free electron LASer in Hamburg), and SACLA (the SPring-8 Angstrom Compact Free Electron Laser) in Aioi is beginning an entirely new era for biological imaging. The intense, ultrashort pulses of X-rays that are available from such free electron laser sources now permit diffraction imaging to be obtained of very small biological structures on timescales that are too short for radiation-induced damage to occur. Conventionally, X-rays from synchrotron storage rings have, over the years, provided invaluable structural information leading to new insights into a whole gamut of biological materials. However, all such measurements have required the sample that is to be probed to be in the form of reasonably large crystals: millions of molecules need to be aligned in a crystal in order for experimentalists to obtain sufficient X-ray signal to produce sensible diffraction patterns. However, there are many classes of biological materials that "refuse" to crystallize. One example of many are the human cell membrane proteins, of critical importance as drug targets. In a new generation of experiments that are beginning to be reported from free electron lasers, structure determinations have now begun to be made of macromolecular material in *nanocrystalline* form; such materials have, thus far, proved difficult to grow as large crystals. As discussed in the following, ultrashort beams of X-rays now enable millions of diffraction patterns to be obtained from an intersecting stream of nanocrystals. Amongst the first biological entities to be studied in this fashion was the membrane protein, photosystem I, in the form of a stream of nanocrystals of dimensions in the range $\sim$200–2000 nm.

It is known that membrane proteins play a critically important role in determining the way cells and viruses function. However, insights into their structure and the dynamics that drive their functioning continue to remain elusive. Photosystem I is one of the largest of membrane protein complexes (it has a molecular mass of 1 MDa; it comprises as many as three dozen proteins and 381 cofactors). Acting as a biosolar energy converter, photosystem I is at the heart of the process of oxygenic photosynthesis (Chap. 7). One of the difficulties of gaining information about this membrane protein has been its intransigence in producing large enough crystals for experiments to be carried out. Crystals of photosystem I are expected to possess symmetry of space group $P63$, with unit-cell parameters $a = b = 281$ Å and $c =$ 165 Å; typically, each crystal comprises as much as 78 % solvent by volume.

The major factor that has, for a long time, limited the attainable resolution in biological imaging has been radiation damage when energetic electrons or X-rays are used [3]. The ultrafast X-ray pulses that are becoming available from FELs circumvent radiation damage by using single X-ray pulses that are of such short duration (10 fs, and even lower) that the pulse has disappeared well before there is the possibility of any damage to the irradiated sample manifesting itself [4]. This was first demonstrated at experiments carried out at FLASH: "diffraction before destruction" at resolution lengths down to 60 Å was demonstrated on samples fixed on silicon nitride films [5]. Subsequent experiments on photosystem I, carried out at LCLS, demonstrated that the concept of "diffraction before destruction" can also be operative at sub-nm resolution. This enabled the establishment of a protocol for structure determination based on accumulating X-ray diffraction data from a sample that takes the form of a stream of photosystem I nanocrystals [6]. It is interesting to note that these nanocrystals were no larger than 300 nm along their edge, that is equivalent to only 10 unit cells!

For those biological entities that are intrinsically non-crystalline, it has been possible to use X-ray pulses from an FEL to obtain diffraction images of materials like the mimivirus by injecting a beam of cooled mimivirus particles into the X-ray beam [7]. Mimivirus is a microbe-mimicking virus, which is among the largest of known viruses [8]; its large size (about 0.75 μm in diameter) makes it more or less comparable to the size of the smallest living cells. This makes it too large for a full three-dimensional reconstruction by means of conventional cryo-electron microscopy. Results of calculations carried out concomitantly with the FEL experiments offered indications that the energy deposited into the virus by the X-ray pulse heated the virus to a temperature of over 100,000 K after the pulse has left the sample, once again demonstrating the efficacy of extremely bright, coherent X-ray pulses being so ultrashort that key damage processes simply do not have time to occur before a diffraction pattern is recorded from a large macromolecule, a cell, or even a virus prior to it becoming plasma and exploding.

As in the case of experiments with photosystem I there was no measurable deleterious effects of the X-rays on the sample. It has been conjectured that X-ray pulses from the FEL induce hydrodynamic explosion of μm-sized objects such that the outside portion of the irradiated sample "burns" first, with heat being transported inwards, rarefying and destroying outer contours first. As a consequence, trapped electrons move inwards into the sample, neutralizing the core which has become positively charged. The inward movement of electrons leads to a positively charged outer layer, which then peels off on picosecond timescales [5].

It is anticipated that considerable progress will be made in the coming years in utilizing ultrafast X-rays from FEL sources in order to determine in vivo crystal structures. Both prokaryotic as well as eukaryotic cells contain proteins in microcrystalline form. Such macromolecules have, almost always hitherto, been studied by first biochemically extracting them from the cells and then recrystallizing them in vitro for conventional X-ray crystallography, usually with synchrotron sources. Cell biologists have long pondered over whether the in vitro macromolecular structures that are experimentally determined fully resemble the real in vivo structures. Femtosecond crystallography that is now becoming accessible at FEL sources makes the question answerable, and the earliest experiments addressing this issue have begun to be performed [9]. Experiments have been conducted at LCLS on a toxin (Cry3A) that is produced by *Bacillus thuringiensis* (Bt) and stored in bacterial cells as naturally-formed microcrystals. The measurements involved the streaming of whole Bt cells into the X-ray beam from the LCLS FEL. The photon energy in the beam was ~9 keV, corresponding to ~1.5 Å, with a pulse duration of 40 fs, a repetition rate of 120 Hz, and a flux of $\sim 10^{11}$ photons per pulse. The naturally grown crystals of Cry3A in Bt cells were injected into the path of the X-ray beam; they traversed a distance of

~200 μm, taking about 3.0–7.5 μs to reach the X-ray beam. The conjecture was that even if this time is sufficient for the cell walls to rupture—an unlikely scenario—it is still not sufficient for the Cry3A crystals to dislodge from within the Bt cells; there would most certainly not be enough time for the toxin molecules of the cellular crystals dislodged from Bt to recrystallize into some different form. As it turns out in this particular case, the structure that was determined for the in vivo crystals was essentially the same as that for in vitro crystals. Structure of in vivo crystals have also been determined in the case of glycosylated cathepsin from *Trypanosoma brucei* (TbCatB) [10], knockdown of which has been shown to be lethal to the parasite responsible for sleeping sickness that affects several million people across Central Africa. The enormous potential that femtosecond X-ray diffraction offers for cell biology cannot be understated. As already noted, considerable progress is anticipated in this emerging area of ultrafast cell biophotonics, perhaps even the prospect of addressing one of the central themes of cell biology, that cellular components are structured: cells are far more than floppy bags of freely moving molecules.

So, how are femtosecond X-ray pulses generated in an FEL? Unlike conventional femtosecond laser systems that rely on the light-emitting properties of materials for lasing, the heart of an FEL lies in producing high energy electrons and manipulating their trajectories such that the light they emit is at X-ray wavelengths (1–0.1 nm). As it is primarily free electrons from ionization of atoms that are the progenitors of the X-rays, this relatively recent entrant into the world of lasers is referred to as an X-ray free electron laser (XFEL). Figure 11.1 is a schematic depiction of how electrons are produced, by irradiating a photocathode with an intense femtosecond laser beam and, subsequently, how they are accelerated and their trajectories manipulated by means of an undulator so as to produce energetic photons.

The fast electrons pass through the undulator in an uncoordinated manner; consequently, the light that they emit is incoherent. The strategy employed in an XFEL uses

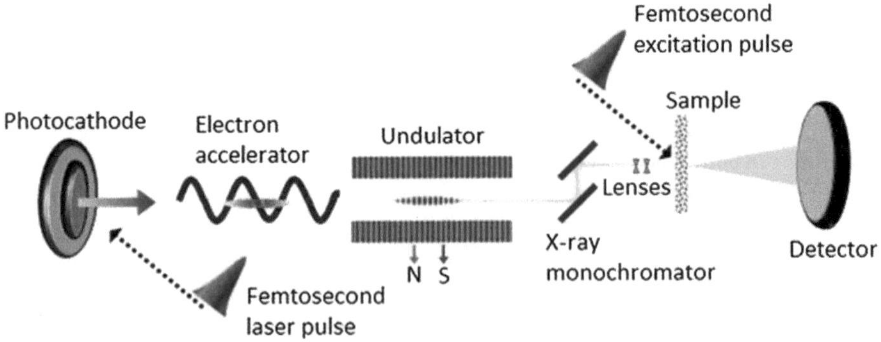

Fig. 11.1 Schematic depiction of how ultrafast X-ray pulses are generated in a free electron laser. An intense femtosecond laser irradiates a photocathode. The emitted electrons are accelerated through an undulator—a region of alternating magnetic fields comprising a series of magnets with North-South...SN...NS...SN configuration. The ensuing acceleration results in formation of ultrashort X-ray pulses which are made to impinge on a biosample. This sample can be simultaneously excited by a second femtosecond laser beam and the X-rays interrogate the excited biosample on ultrashort timescales

the concept of self-amplified spontaneous emission (SASE). Instead of the energetic electrons oscillating through the resonator many times so as to increase the effective interaction length (as in a conventional laser cavity), in SASE the undulator is timed such that light emitted by a bunch of electrons interferes with the preceding electron bunch. This photon-electron interaction results in temporal and spatial modulation of electron density, generating spectrally narrow emission with relatively long coherence length. SASE has been successfully employed in all XFEL facilities to create useable beams of coherent X-rays.

Although the use of femtosecond-duration, coherent X-ray beams from XFEL facilities are opening new vistas for diffraction imaging that unveils the internal structure of cells and organelles, it is pertinent to draw brief attention to on-going experimental difficulties that still need to be resolved. One concerns the attainable spatial resolution: although very good, it is, nevertheless, limited to a few tens of nanometers. This is a consequence of the poor scattering power of most biological samples and, hence, the continuing challenge for experimentalists is to devise new methodologies that will help recover correct phase information from measured diffraction images that inevitably possess low signal-to-noise ratio and from lowest-resolution data.

Some novel approaches are beginning to be explored in attempts to enhance the attainable spatial resolution by (i) seeking to enhance diffraction signals and (ii) by more robust phasing. The weak diffraction signals that emanate from most biological objects have recently been sought to be enhanced by interference with strong diffracting waves from dispersed colloidal gold particles that are embedded within the biological sample of interest. X-ray imaging is carried out of these Au-particles and the biological object of interest together [11]. Interference occurs between the strong diffraction waves emanating from the Au-particles on the one hand and the much weaker diffraction waves from the biological object on the other. Such interference is being utilized to enhance the signals from the biological object to a much more detectable level by the process of coherent amplification (see Sect. 11.3). Experimentally, the easily-observed positions on the detector of the Au-particles, as determined by conventional Patterson analysis [11], serve as the initial phase. As Au-particles are relatively unreactive, this approach is likely to prove to be compatible with the X-ray imaging of a wide variety of biological objects under physiologically-relevant conditions; early demonstrations from experimental images obtained using SACLA appear to testify to this [12].

The next major XFEL facility is expected to be the European XFEL, the precursor of which is FLASH. Table 11.1 summarizes some of the key attributes of this next-generation XFEL and compares it with the attributes currently exhibited by SACLA and LCLS.

It is also of interest to make comparison of X-ray photons from an FEL with photons from femtosecond infrared lasers. As the ponderomotive energy (Chap. 2) gained by ionized electrons depends on the inverse square root of photon energy, it is clear that processes like above-threshold ionization, rescattering, and its consequences, such as high harmonic generation, that are readily encountered in femtosecond duration pulses from a conventional laser (Chap. 2) are essentially "switched off"

Table 11.1 Comparison of the attributes of XFEL facilities: SACLA in Aioi, LCLS in Stanford, and the forthcoming European XFEL in Hamburg

Property	SACLA	LCLS	Euro-XFEL
Commissioning date	2011	2009	2017
Repetition rate	60 Hz	120 Hz	27 kHz
Shortest wavelength	0.08 nm	0.15 nm	0.05 nm
Max. electron energy	8 GeV	14.3 GeV	17.5 GeV
Length	0.75 km	3 km	3.4 km
No. of undulators	3	1	3–5
Peak brilliance[a]	1×10^{33}	2×10^{33}	5×10^{33}
Average brilliance[a]	1.5×10^{23}	2.4×10^{22}	1.6×10^{25}

[a] In units of photons/s/mm^2/mrad^2/0.1 % bandwidth

in the case of photons from an FEL. For 92.5 eV photons that are readily available from, for instance, FLASH, at an intensity of 10 TW cm^{-2}, the corresponding photon wavelength is 13.4 nm and, hence, the ponderomotive potential is only 160 meV. In contrast, with a photon pulses of similar peak intensity from a Ti:sapphire laser, where the photon energy is 1.55 eV, the ponderomotive potential is a massive 600 eV. For Xe atoms—whose first ionization energy is 12.13 eV, the FLASH parameters yield a Keldysh parameter, γ, of 8.7, well into the multiphoton ionization regime. Hence, FLASH (as well as LCLS and SACLA) photons of 10 TW cm^{-2} peak intensity will not induce tunnel ionization while with photon pulses of the same peak intensity from a Ti:sapphire laser the ionization of Xe will be almost entirely due to tunneling.

11.3 X-Ray Scattering Interferometry and Molecular Rulers

Imaging methodologies in cell biology ultimately require both ultrafast time resolution and atomic-scale spatial resolution so that structural changes may be observed in an individual biomolecule [4, 13–16]. In the previous section, we have discussed the ultrafast pulsed X-ray source—the XFEL—which can provide the required time resolution. In the visible range, in vivo observations have significantly progressed due to remarkable developments in labeling and single-molecule fluorescence techniques [17, 18]. Appropriate labeling in single-molecule fluorescence measurements now succeeds in providing positional information with an accuracy of $\sim \frac{\lambda}{100}$, far better than the optical diffraction limit $\sim \frac{\lambda}{2}$. It would be desirable to achieve similar $\sim \frac{\lambda}{100}$, picometer-scale resolution [19, 20] in X-ray imaging for determining structures of biomolecules and, indeed, for dynamical studies.

The distances between different molecules or moieties in a given macromolecule are closely related to its 3D structure. Hence, for a macromolecule with a dynamic conformation, distributions of distances between many different pairs of molecules or moieties can, in conjunction with a suitable theoretical model, help define the structural dynamics. Thus, in principle, the separation between two points—a "molecular ruler"—may be sufficient to experimentally deduce the molecular structure [21–23]. Though this is a promising approach, most existing rulers are sensitive only to relative changes in distances. They do not provide absolute distances; nor do they provide accurate enough occupancy distributions, particularly for a complex system in which multiple distances (or conformations) coexist. These limitations arise from (i) averaging of signals over a finite time, (ii) complex nonlinear and non-unique mapping between the observed signal and the molecular dynamics, and (iii) possible contributions to the observed signal from other parameters apart from distance. These limitations result in the lack of distance calibration on an absolute scale, preventing quantitative comparison of experimental results with those obtained from computational models [21–23].

Small-angle X-ray scattering interferometry address these issues and provides instantaneous and high-precision distance information that enables precise determination of macromolecular structures [21–23]. As in the case of fluorescence tagging [17, 18], biomolecules and proteins, for example DNA, are tagged with gold nanocrystals. Two gold nanocrystal probes are attached to two specific locations in a macromolecule [Fig. 11.2 (top)], and the mutual interference in the X-ray scattering

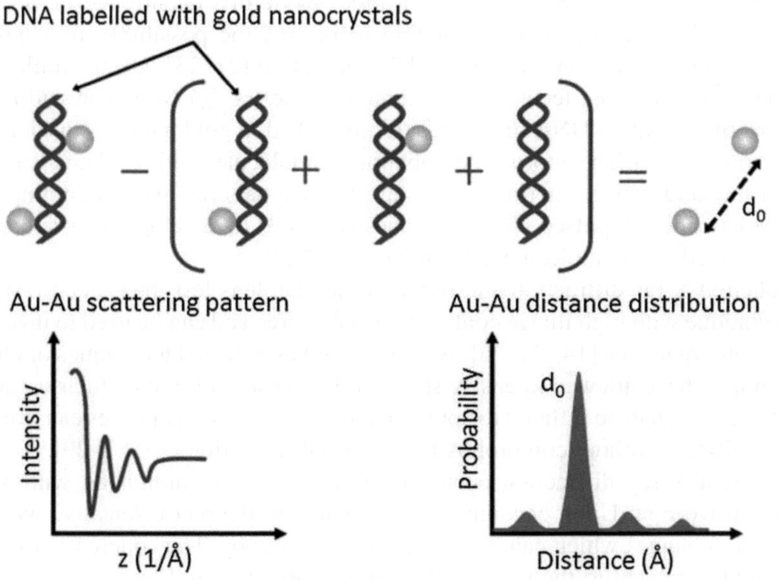

Fig. 11.2 Schematic of X-ray scattering interferometry and a "molecular ruler". A double helix DNA is labeled with a gold nanocrystal on each of the two DNA strands (*top*). The X-ray interference pattern, on Fourier transforming after subtracting the scattering signals from individual gold nanocrystals and the DNA molecule, yields information about the separation between the two nanocrystals (*bottom*)

of gold nanocystals is measured. As the scattering from gold nanocystals is much stronger than from organic molecules, due to the higher atomic number of gold and because the scattering is "instantaneous" compared to the slow atomic motions of the molecule, the interference pattern needs a much shorter time to record. Moreover, it provides precise information about the relative distance between the two nanocrystals. Fourier transforming the recorded interference pattern after necessary data processing (Fig. 11.2) directly provides an un-averaged snapshot of intramolecular distance between the gold probes, a structural measure of the thermodynamic landscape of conformational states. A major advantage of this technique is that data analysis is totally model-independent and does not require a priori knowledge (or conjecture) about the structure [21–23].

X-ray scattering interferometry on DNA double helices has been reported in which gold nanocrystal probes were placed at 18 different pairs of positions [21–23]. Distance distributions were measured to determine structural values for DNA helices in solution for comparison with other measurements and theoretical models. The results highlighted the limitations of the prevailing understanding that DNA behaves like a stiff elastic rod with respect to inherently ubiquitous mechanical deformations. Experimental investigations have revealed that in the absence of applied tension, DNA is at least one order of magnitude softer than indicated by single-molecule stretching experiments. Results have also shown a quadratic dependence between the end-to-end length in contrast to the linear dependence predicted by the conventional elastic rod model, suggesting more flexibility, occurrence of stretching over more than two turns of the double helix, and the possibility of long-range allosteric communication through the DNA structure [21–23]. Using similar labeling and X-ray tracking techniques, picometer-scale ($\sim\frac{\lambda}{100}$) Brownian and rotating motions of individual DNA molecules tightly linked to gold nanocrystals have also been reported [19]. The approach of labeling biomolecules with gold nanostructures and of tracking their scattering and/or interference patterns may, in coming years, also be extended to pulsed electron probes (Chap. 3) to develop a laboratory-scale time-resolved single-molecule detection system [20].

Coherent X-ray diffraction microscopy is another lens-less, phase-contrast imaging technique with high image contrast; it is label-free and can be used to investigate single cell organelles [14, 24–26]. Although well established techniques of electron tomography have allowed intensive study of the 3D structure of cellular organelles, they have proved to be difficult to apply to thick samples. X-ray probes can overcome these limitation without compromising on spatial resolution [25, 27–29].

Coherent X-ray diffraction occurs when a sample is illuminated with a well-defined wavefront. Upon encountering the scatterer, the phase-sensitive wavefront undergoes changes which can be utilized for microscopy. The coherently diffracted wavefront is related to the sample electron-density map by Fourier transform, but the phase of the diffracted wave is not directly measurable unless more complex interferometric techniques are employed. Moreover, phase information has to be retrieved for complete 3D reconstruction of the image. As already noted, coherent X-ray diffraction microscopy provides phase contrast information without using lenses or sample tagging [25, 30]. Thus, as in optical microscopy, it is superior to conventional absorption contrast imaging X-ray microscopy [28, 29], particularly for transparent objects like human chromosomes and DNA [25, 31]. Being sensitive, it also offers advantages in terms of lower radiation dose and better signal-to-noise ratio (faster data collection) and higher image contrast, even in an aqueous environment. Reconstructed 3D images of a thick biological sample—an unstained human chromosome—have successfully reproduced an axial structure which other microscopic methods have been unable to visualize without staining [25]. The concept of coherent X-ray diffraction can be further extended to perform Talbot interferometry using a pair of diffraction gratings [32–34]. In this case the phase object is placed in front of one of the gratings and its structural information is obtained by analyzing the deformation of the fringe pattern. Using this technique, in situ properties of transparent biological specimens like eye lenses have been measured [32]. The eye lens, a major ocular component, has a gradient refractive index. The refractive index variations across the lens are dependent on the distributions and concentrations of various proteins and, consequently, of various electron density patterns. The interferograms map these variations and help reveal previously undetected conformations which might be linked to various developmental stages [32].

Though most of these techniques currently employ a continuous wave X-ray source, they can all be adapted for use with ultrafast pulsed X-ray and/or electron sources, which will enable them to simultaneously provide high spatial as well as temporal resolution. These techniques will then be able to shed light in hitherto unexplored directions in ultrafast biophotonics by correlating structures to mechanisms and functionalities of biomolecules and proteins.

11.4 Terahertz Spectroscopic Probes

Lab-on-a-chip DNA analyzers, DNA biosensors, and gene chips constitute contemporary techniques that continue to revolutionize genetic diagnostics. They mostly identify polynucleotide sequences by detecting the binding of unidentified DNA molecules to single-stranded oligo- or poly-nucleotide DNA of known base sequence. The unidentified DNA are the "target" while molecules with known sequences are the "probe", and the basis of these techniques is the fact that binding of target to probe molecules is most likely to occur when they possess complementary base sequences. Such binding is referred to in the literature as "hybridization" and its detection currently relies on fluorescent labelling (and sensitive optical detection) of molecules of target DNA [35].

It has long been accepted that such labeling can often be accompanied by certain deficiencies, such as the fluorescence efficiency showing dependency on the binding site, and fluorescence yields being variable [36, 37]. From an operational point of view, labelling often appears to be a step that is unwanted by practicing biologists in that it appears to introduce complications to the analytical procedure for gene detection. Consequently, it is not surprising that there has long been an interest in seeking an analytical technique that does away with labelling and it is in this context that there is contemporary interest in exploring terahertz (THz) methods.

THz radiation covers the part of the electromagnetic spectrum that spans the frequency range from about 0.1–10 THz. This range is especially appropriate for probing rovibrational excitation in molecules, both neutral and in ionized form, as well as localized vibrations that often accompany conformational changes in DNA. THz imaging offers several advantages that optical, nuclear magnetic resonance and X-ray methods do not, especially relating to conformational information about biomolecules in cells and tissues and how such changes might relate to biological functions [38]. One shortcoming of THz radiation in the context of biology is its very strong absorption by water, a medium in which most biological entities are immersed. Consequently, most contemporary diagnostic work that employs THz radiation has been carried out on dry samples, often in powder or thin film form. Studies of THz absorption have been reported on some proteins, nucleic acids, and tissues [39–44]. Very recently, time-domain spectroscopy in the THz region has begun to be successfully applied to enable studies to be carried out on biological samples in aqueous media [45]. Label-free detection of single-base mutations on oligonucleotides in aqueous solutions has been reported [45] in which a Ti:sapphire laser is utilized to produce sub-100 fs pulses that irradiate an InAs emitter of THz radiation. After passing through DNA molecules in aqueous phase, the THz field is detected using electro-optical sampling with a ZnTe crystal. Such experiments can be conducted at room temperature but stringent humidity control is mandatory (typically, relative humidity levels need to be kept below about 5 %). As can be seen in Fig. 11.3, observable changes in absorption spectra that can be ascribable

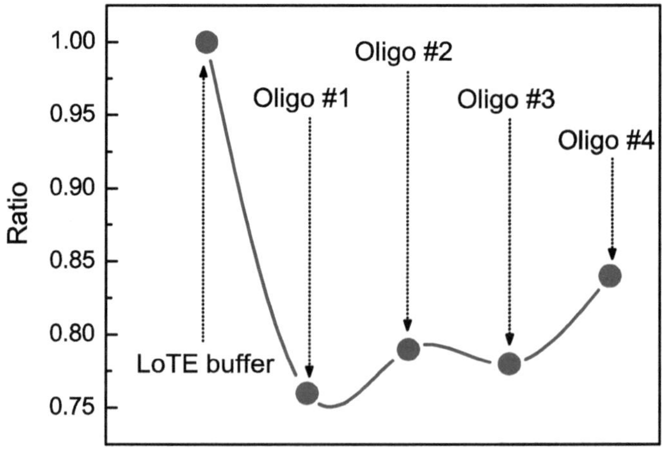

Fig. 11.3 Ratio of the THz signal being transmitted through neat LoTE buffer and LoTE buffer containing different oligonucleotides, Oligo #1–Oligo #4, based on measurements conducted by Tang et al. [45]. Peaks in the THz transmission function for neat buffer and oligonucleotides occur at 0.59–0.88 THz and the ratio that is plotted is the transmitted signal intensity: I at 0.88 THz/I at 0.59 THz

to single-base mutations have been convincingly demonstrated over the frequency range 0.2–2.6 THz [45].

The oligonucleotide sequences with single-base mutations referred to in Fig. 11.3 are as follows:

Oligo #1: TGGAGCTGGT***G***GCGTAGGCA;
Oligo #2: TGGAGCTGGT***C***GCGTAGGCA;
Oligo #3: TGGAGCTGGT***T***GCGTAGGCA; and
Oligo #4: TGGAGCTGGT***A***GCGTAGGCA.

The bold, italic letter in each oligonucleotide delineates the different bases. As is seen, single-base mutations give rise to (i) significantly lower THz transmission when compared to the neat buffer solution and (ii) the four oligonucleotides exhibit different absorption spectra, at least over the frequency range 0.2–2.6 THz. The relative changes in absorption peaks hold much promise for useful applications in future for THz spectroscopic methods of detection, in label-free fashion of single-base mutations.

11.5 Frequency Comb Fourier Transform Spectroscopy

Along with THz, the mid-infrared spectral region of 2–20 μm (500–5000 cm^{-1}) is a domain of very considerable interest in biophotonics because a large number of biomolecules exhibit strong characteristic vibrational transitions in this spectral range. Consequently, mid-infrared spectroscopy has proved to be a powerful tool for understanding the molecular structure governing microscopic mechanisms, and for performing non-intrusive diagnostics of composite biological systems [46, 47]. The low Rayleigh scattering losses in this spectral range are advantageous for tomography and imaging in turbid biological media. Fourier Transform Infrared (FTIR) Spectroscopy has emerged as one of the major tools for biophotonic and biomedical applications and is making significant contributions in clinical evaluation by providing molecular-level information that allows investigation of functional groups, bonding types, molecular conformations and surrounding environment. Research has been carried out on a number of natural tissues using FTIR spectroscopy because it is a relatively simple, reproducible, and nondestructive technique requiring only a small amount of sample with minimum preparation time [46–48].

However, identifying multiple, and possibly unknown, biomolecules requires a detailed spectroscopic analysis over a broad spectral bandwidth with high spectral resolution. A broad frequency "comb" comprising a series of discrete, equally spaced frequencies might prove to be an ideal light source for such applications [46, 49–55]. Laser frequency combs are coherent light sources that emit a broad spectrum of discrete, evenly-spaced narrow lines whose absolute frequency is stabilized and measured to within the accuracy of an atomic clock without the need to employ a conventional dispersive element-based spectrometer. They can be generated by a number of mechanisms, including amplitude modulation of a continuous wave

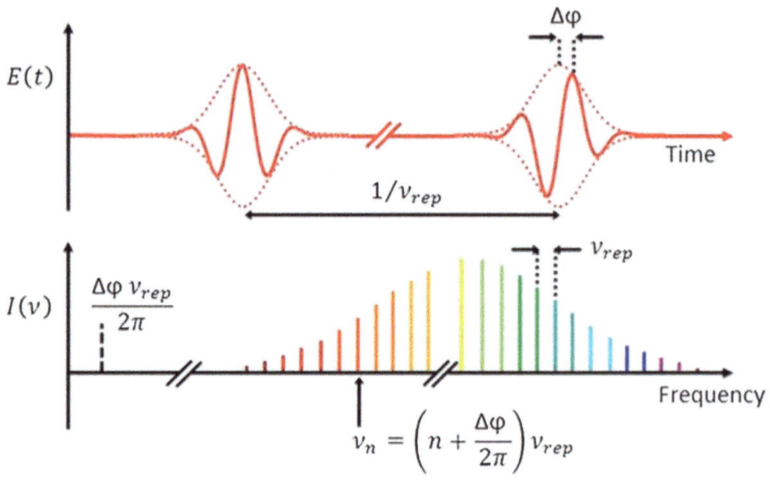

Fig. 11.4 Time (*top*) and frequency (*bottom*) domain representations of an ultrashort pulse train emitted by a femtosecond laser. The spectral width of the frequency comb is inversely proportional to the pulse duration

laser or stabilization of the pulse train emitted by a mode-locked femtosecond laser. Their development in the near-infrared and visible domains has revolutionized optical frequency metrology [46, 49, 50]. Attempts are now being made to extend frequency comb techniques to the mid-infrared spectral region for FTIR and vibrational spectroscopy. Efficient mid-infrared frequency comb generation using ultrafast laser sources promises to significantly expand the variety of possible applications even further [46, 56, 57]. In particular, new spectroscopic and imaging techniques in the "fingerprint region" with dramatically improved precision, sensitivity, recording time and/or spectral bandwidth are being developed which may provide new insights into several unexplored topics in ultrafast biophotonics.

As schematically depicted in Fig. 11.4, the periodic pulse train of a mode-locked femtosecond laser can give rise to a regular comb spectrum of a large number of laser modes with a spacing equal to the pulse repetition frequency [46, 49, 58–60]. Every time the ultrashort pulse circulating in the laser cavity reaches the output coupler, a fraction of its energy is coupled out of the laser. The emitted pulse train therefore has a repetition frequency (ν_{rep}) given by the inverse round-trip time of the pulse in the laser cavity. Due to dispersion in the laser cavity, the carrier wave of the pulse propagates at a phase velocity that differs from the group velocity. Therefore, the electric field of the out-coupled pulses is shifted with respect to the pulse envelope by an amount $\Delta\varphi$ (Fig. 11.4). The Fourier spectrum of the electric field of such a pulse train comprises a series of sharp, equispaced frequency components given by $\nu_n = (n + \frac{\Delta\varphi}{2\pi})\nu_{rep}$, where n is an integer. For applications in precision metrology and spectroscopy, it is mandatory to stabilize $\Delta\varphi$ with high accuracy as any jitter in $\Delta\varphi$ results in broadening of the comb linewidth. There are modelocked femtosecond lasers which can routinely generate phase stabilized pulse trains with few cycle pulses. Thus commercially available broadband frequency combs have now matured to become standard instruments of precision metrology. The extension of high-sensitivity frequency comb techniques to new spectral regions, from THz frequencies to the extreme-ultraviolet, is likely to open new horizons for advanced diagnostic instruments in biomedicine [46, 49, 58–60]. Several frequency comb generation schemes involving mode-locked lasers, difference frequency generators, optical parametric oscillators, and micro-resonator based Kerr-comb generators have been proposed and demonstrated [46, 56, 57]. The last one in particular also has a small chip-sized footprint and can, therefore, be easily integrated into handheld or portable devices.

Novel techniques have recently been demonstrated in which the frequency comb can be directly used to investigate multiple transitions in an absorbing sample [46, 51, 53, 61–65]. They have been most successfully applied to mid-infrared spectroscopy of gaseous samples which have strong molecular fingerprints but the methods are also promising for liquid and solid samples [66]. These techniques can potentially provide short measurement times (<1 μs), high sensitivity (<1 pm), and high accuracy over a broad spectral bandwidth compared to conventional spectroscopy [64]. Experimentally, the frequency comb is selectively attenuated and phase-shifted by molecular resonances; analysis of the latter relies on well-established Fourier transform spectroscopy. The spectroscopic signal is recorded by a single photodetector, overcoming the limitations of dispersive-element based spectrometers [46, 67]. Two different implementations of Fourier transform spectrometers have been reported: Michelson-based Fourier transform spectroscopy, and dual-comb spectroscopy. Each has its own distinct advantages, but both rely on the down-conversion of the high comb frequency signal to a lower frequency range that can be electronically processed. Both schemes are an example of a time-domain technique in which the pulse train of a comb is interferometrically sampled—like in a sampling oscilloscope—by a second pulse train of different repetition frequency [46, 67]. The two schemes shown in Fig. 11.5 differ in the technique used for generating the second pulse train: a Michelson interferometer (Fig. 11.5a) or a second synchronized frequency comb source (Fig. 11.5b). The two interfering frequency combs exiting the interferometer pass the absorbing sample and beat on a photodetector to generate a time-varying low frequency signal [46, 67].

Michelson-based Fourier transform spectroscopy employing an incoherent or coherent broadband source is a well-established technique [46–48]. Therefore, replacing the traditional light source with a frequency comb is not difficult. Dual-comb spectroscopy, in contrast, has only recently been introduced and its full potential is yet to be explored. Use of a mid-infrared frequency comb dramatically reduces measurement time and improves signal-to-noise ratio, even in single-shot measurements, owing to the high spectral radiance of such a coherent source. The sensitivity may be further enhanced by increasing the path length—perhaps by employing a multipass cell or a high-finesse resonator [46, 51, 53, 61–65]. Extremely high sensitivity (one part per billion) has been demonstrated using Michelson based mid-infrared comb FTIR spectroscopy of H_2O_2 [46, 68]. Another intriguing application explores

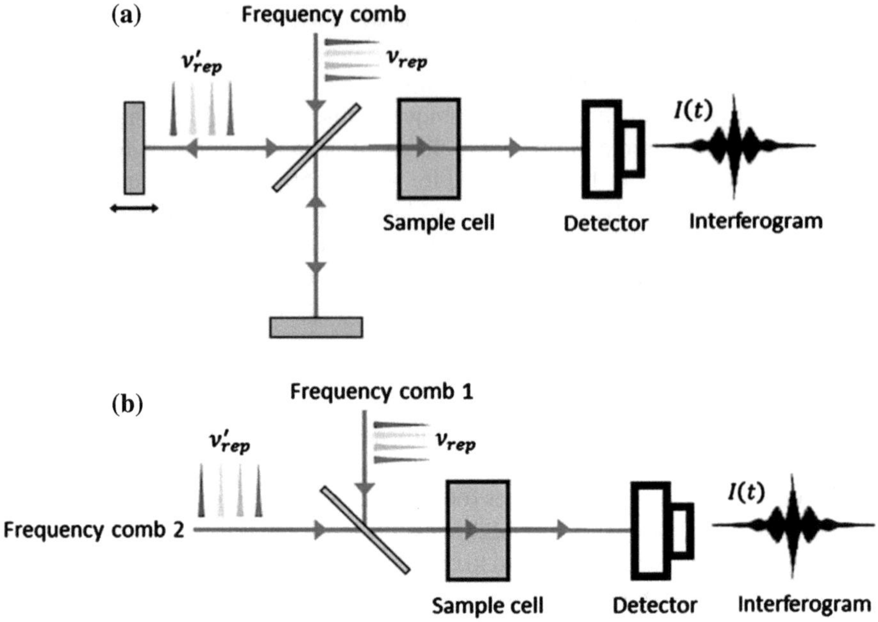

Fig. 11.5 Michelson-based (**a**) and dual-comb (**b**) Fourier transform spectroscopy. Two synchronized pulse trains with slightly different repetition frequencies are mixed to generate a cross-correlation interferogram $I(t)$

combining the potential of broadband near-field microscopy with a frequency comb to interferometrically map the near-field interaction between a metal tip and polar organic molecules [69].

An important practical advantage of a dual comb approach is that being a static scheme, it overcomes shortcomings induced by the moving mirror of the Michelson interferometer; shorter measurement times are the result [46, 52, 55], with spectral resolution being limited only by the measurement time and the intrinsic comb linewidth. The lack of efficient mid-infrared frequency comb sources and the difficulty associated with synchronizing the pulse trains of two combs with interferometric precision have hindered development of this technique. However, an exceptionally high spectral resolution of 12 GHz within a measurement time as low as 10 μs has been demonstrated in proof-of-the-principle experiments [46, 70]. Coherent stimulated Raman dual-comb (imaging) spectroscopy is an alternative technique for accessing fundamental transitions with near-infrared combs [71]. As was done for Michelson based techniques, dual comb based scanning near-field optical microscopy have also been explored [72]. Like in the case of conventional spectroscopy, the spectral range in frequency comb spectrometers is also governed by the bandwidth of the source. However, in frequency comb techniques, unlike in the case of dispersive-element based spectroscopy, the resolution and speed are also governed by source parameters [46, 51, 53, 61–65].

Though frequency comb spectroscopy is still in its infancy, combining it with other laser-based tools in nonlinear optics and electronic signal processing may vastly enhance its range and capabilities. One may even envision chip-size comb spectrometers based on microresonators being developed for real-time biomedical applications. Other than precision spectroscopy, frequency combs may also be used for line-by-line pulse shaping [73], to achieve coherent control over molecular vibrational excitations in biomolecules. Traditional pulse shaping techniques rely on optical elements that are either not transparent or exhibit very low diffraction efficiency in this spectral region, limiting the modulation of the time varying electric field. Many other interesting biophotonic applications, like frequency comb optical coherence tomography [46, 74] of cells and tissues are also likely to emerge in the near future.

11.6 Coherent Control of Biomolecules

Coherent control is a technique for controlling dynamical processes by means of a tailored external optical field. It is based on the quantum mechanical principle that upon excitation, a system can be found in a state that is a superposition of two or more states. The aim here, as schematically depicted in Fig. 11.6, is to steer quantum superposition or interference of the initial states to a specific target state via an appropriately chosen optical field [75–77]. Generally, the light field has to be "shaped" to yield the desired time-dependent spectrum, along with time-dependent phase and polarization to implement coherent control. As ultrashort pulses have a broad spectrum, they permit a wide range of possibilities. The simplest shaped pulse is a chirped pulse, with a time varying frequency (Chap. 2). Accordingly, strong chirp

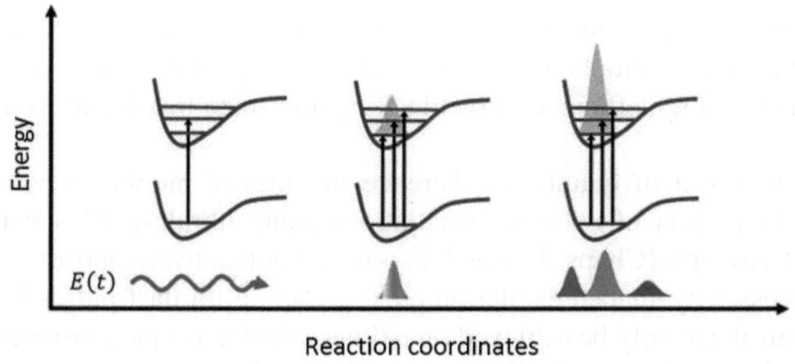

Fig. 11.6 Schematic representation of coherent control. *Left* A cw excitation generally excites only a single vibronic state of the system. *Middle* A transform-limited short pulse excitation results in simultaneous excitation of several vibronic states. The quantum interference between these states generates a wavepacket. *Right* A shaped- or chirped-pulse comprising frequency components with time-varying amplitudes, phase and polarization selectively excites vibronic levels sequentially, each with different efficiency. Such a mechanism affects the interference among several pathways, resulting in enhanced outcome for a particular channel. Thus, proper optimization of the pulse sequence enables control of the photoreaction in a coherent fashion

dependence has been demonstrated in several light-matter interactions like ionization [78, 79], high harmonic generation [80] and nonlinear Raman spectroscopy [81]. Along with a shaped pulse, the technique also needs an automatic feedback control to optimize the yield of the target state [82]. Efficient pulse shaping techniques based on spatial light modulators [83, 84] and optimization control routines involving evolutionary algorithms have been developed [85, 86]. The computational task of finding a control field for a particular target state is difficult and becomes even more difficult with increase in the size of the system. Therefore, it is often possible to control only the outcome of a given chemical reaction, usually involving a specific molecule. In spite of complexity imposed limitations, the technique has diversified, being used in spectroscopy [87–89], quantum information processing [90], laser cooling [91, 92], and plasmonics [93]. The questions to be addressed here are: can coherent control be implemented on photoreactions that are relevant to ultrafast biophotonics? Is it possible to steer biologically important reactions by designing the temporal structure of an optical field?

A photoreaction, similar to that occurring during photosynthesis or vision (Chaps. 7 and 8), involves multiple initial states and can lead to more than one product state. Since these photoreactions are triggered by sunlight or light from incoherent sources, they cannot be steered by the temporal structure of the incident light field. Also, in complex protein molecules, random fluctuations among the enormous number of degrees of freedom might be expected to cancel any interference effects. At the same time, as we have seen in Chaps. 7–9, proteins are highly evolved structures. Indeed, nature steers relevant photoreactions into the desired final states by employing coherent superposition of the protein states as well as those of the surroundings. The quantum interference between multiple pathways in biological systems yields very high efficiency and adaptivity. In the context of "externally-shaped-excitations", coherent control might not be considered to be applicable to biological systems in physiological surroundings. However, its implementation, particularly on a single-molecule level (Chap. 4) could be used as a tool to gain deeper insight into coherent processes in ultrafast biophotonics. It might also help in designing adaptive artificial systems with high efficiencies by blocking unwanted transitions as in biological systems.

An alternate test of quantum coherence in ultrafast biophotonics—other than real-time observation of coherent oscillations using ultrafast 2D spectroscopy or electron microscopy (Chaps. 3, 4 and 7)—is to photoactivate and manipulate biological processes by shaped excitation pulses. Use of ultrafast pulses is mandatory because control can only be achieved on a timescale that is much faster than random thermal motions ($\leq$1 ps), which scramble phase relationships and cause decoherence. The first experiment that used shaped excitation on a biological chromophore involved chirped excitation of the green fluorescent protein (GFP), where adapting the linear chirp to the wavepacket motion in the excited state induced a dynamic Stokes shift [94]. Later, using feedback optimized coherent control of energy flow pathways in the light-harvesting antenna complex LH2 of a photosynthetic purple bacterium, successful steering of $\sim$30 % of the excitation energy into one of the two available product channels was achieved [95]. It is the efficient Raman scattering of chromophores [96] that accounts for the relatively easy convergence of optimization routines and, consequently, the simplified control.

Which other biological systems or biomolecules are promising candidates for coherent control experiments? Of course, all biological proteins and DNA are potential candidates because of the relevance of their folding processes and technological implications like molecular electronics (Chap. 10). Optical control of the simple photochemical reaction responsible for the primary step of photoisomerization of the retinal molecule in bacteriorhodopsin from the all-*trans* to the 13-*cis* state is another example of the application of coherent control in ultrafast biophotonics [97]. It was demonstrated under weak field conditions (where only 1 of 300 retinal molecules absorbs a photon during the excitation cycle) that are relevant to understanding biological processes. By modulating the phases and amplitudes of the spectral components in the photoexcitation pulse, it was shown that the absolute quantity of the 13-*cis* component formed upon excitation could be enhanced or suppressed by $\sim$20 % of the yield observed using a short transform-limited excitation pulse having the same energy. The observations suggested that the shaped-excitation steered the isomerization through interference effects, a mechanism supported by vibrational coherence discussed in Chaps. 4, 7 and 8. A weak field excitation to control the degree of coherence without causing strong perturbation in the biological protein's environment and dynamics is mandatory to understand behaviour under normal functional conditions. Strong field excitation conditions inevitably access higher-lying excited states and substantially perturb the energy landscape through multiphoton processes [95, 97]. Such effects are usually absent under physiological conditions. These issues are important for interpreting the function of proteins in natural surroundings. However, they are irrelevant if the final objective is to enhance a particular reaction outcome. In this context it is apt to point out that strong field control of the optical field variation within a single 2-cycle 800 nm laser pulse (5 fs laser pulses of 10^{15} W cm^{-2} intensity) has been used to selectively break one of the two equivalent bonds in water molecules [98].

11.7 Probing Biology Beyond the Quantum Limit

The sensitivity that can be attained in experiments on biological entities and processes that involve optical visualization and tracking is, of course, ultimately limited by shot noise—an inevitable consequence of the quantization of light. It is, however, becoming obvious that this quantum limit may be circumvented by utilizing optical states with non-classical photon correlations [99]. Non-classically correlated photons, of the type extensively used in physics, for instance in squeezed light experiments [100] and for gravitational wave detection [101], have begun to find utility in several biological contexts, ranging from quantum optical coherence tomography [102] to determining protein concentrations via entangled photons [103]. However, although non-classically correlated photons were used in these (and similar) experiments, the limit imposed by shot noise was still not overcome.

It may, at first sight, be considered unlikely that squeezed light can be of utility in biological environments, the main reason being the use of high numerical aperture lenses (that are the norm in such experiments) used in conjunction with relatively thick biological samples would lead one to assume that the propagating light becomes so distorted that its spatial mode does not match the detection mode. This is one of the constraints that have inhibited the use of squeezed light in biological experiments for a number of years. Recently, they been overcome in elegant fashion [99] by the expedience of a dual-field technique to replace the usual single field acting both as an oscillator and an interrogator of a bio-sample. In this scheme, a Gaussian probe beam interrogates the bio-sample and, thereby, scatters. A second Gaussian beam whose phase is π-shifted, is used to independently define the detection mode. A single photon detector records the resulting interference between these two fields. The utility of this technique for biologically-relevant studies has already been demonstrated [99] with optically trapped yeast cells, *Sacchoromyces cerevisiae.* Intercellular experiments on different cells have established that the thermally-induced motion of lipid granules inside cells are inhibited by a network of actin filaments which cause the cells' thermal motion to be sub-diffusive. In two-field experiments on yeast cells, one optical field was used to optically trap a yeast cell while a second, shaped local oscillator field probed the trapped cells' motion. The parameter that was measured in these experiments was the mean square displacement (MSD) which, for the thermally-induced motion of a free particle, may be expressed as [99]:

$$\Delta x^2(\tau) = \left\langle [x(t) - x(t-\tau)]^2 \right\rangle = 2D\tau^{\alpha}. \tag{11.1}$$

Here, τ is a measure of the time delay between successive measurements and D is the diffusion constant; α is the cell medium's viscoelasticity. In the purely viscous regime, in which each yeast cell will undergo Brownian motion, $\alpha = 1$. The signature of the sub-diffusive regime is when $\alpha < 1$.

Test measurements were conducted on silica beads of a few μm dimension and they yielded a value of α that was unity to all intents and purposes (0.994 ± 0.006) [99]. In measurements conducted on yeast cells, values of α were obtained that ranged from 0.6 to 1.0, depending on which part of the local aqueous environment the lipid particles interacted with. When MSD measurements were made using squeezed light, significantly enhanced precision (by ~22 %) was obtained in comparison with measurements using coherent light, allowing up to 64 % enhancement of measurement rate while, at the same time, maintaining precision. The implication of such experiments is that a significantly better experimental capability appears attainable for making time-dependent measurements of motion in a biological environment. It is certainly looking likely that further use of squeezed light in biological applications will be forthcoming which will enhance the scope of biological phenomena and processes that become amenable to quantitative investigations [104, 105]. As has been shown [103], ultrafast laser science is likely to play an important role in such experiments by permitting microfluidic channels to be laser-written, in materials like fused silica. Writing such microchannels constitutes an extension of well-developed methods of laser-writing of waveguides [106–108], with strong HF acid used to etch the waveguides into hollow channels through which fluids can flow (mainly by capillary action). Crespi et al. [103] have devised an elegant experimental scheme in which a femtosecond-laser-written microfluidic channel is made to pass through one arm of a conventional Mach-Zehnder interferometer. When a liquid biological sample is fed into the microfluidic channel, the relative phase shift of light that occurs in the sensing arm of the interferometer manifests itself in the interference fringes that are observed.

11.8 Probing Biology on Attosecond Timescales

It has been demonstrated that trains of attosecond (as) pulses as well as isolated pulses of duration as short as ~70 as can be reliably generated by means of the high-harmonic generation process. Typical photon energies in attosecond pulses range from ~15–120 eV. Experiments have begun to be conducted that probe dynamics in the attosecond regime, and these have, in the main, used a combination of a single attosecond pulse and a carrier-envelope-phase (CEP) stabilized pulse in the near infrared (~800 nm wavelength) whose pulse duration is typically ~5 fs. This, in itself, presents a not-inconsiderable technical challenge in achieving temporal overlap of ultrashort and not-so-ultrashort pulses so as to achieve attosecond time resolution. Initial experiments on atomic systems have yielded intriguing results but their interpretation continues to present challenges. The influence of *two* intense, time-dependent optical fields has to be unravelled, keeping in mind that molecular potential energy surfaces—of even a one-electron molecule like H_2^+—undergo very substantial morphological changes in the course of an intense, 5 fs long, CEP-stabilized pulse [109, 110]. It is certain that attosecond time-resolved experiments on biomolecules will be rewarding once conceptual barriers (like how to properly deal with the inevitable transient distortion of electronic structures in ultrashort pulses of intense optical fields) are surmounted.

Biological entities that are sought to be probed on attosecond timescales will, almost inevitably, be in the condensed phase. As has been noted [111], such experiments will open the possibility of following, with high sensitivity, changes in the oxidation state, chemical environment and electronic state of the atomic and molecular constituents by monitoring the effect that electrons (or holes) have on various sites within the medium. The use of attosecond excitation in such studies will make it possible to record "images" of biochemical processes like charge migration within diverse media of biological significance.

We have already discussed tunneling effects that occur in biological molecules in Chap. 9. Tunneling processes occur on attosecond and femtosecond timescales. There are intriguing possibilities of tunneling effects proving useful for developing new DNA sequencing methodologies. We know that it is the sequence of four bases—adenine (A), guanine (G), thymine (T), and cytosine (C)—that determines genetic information which governs the structure and function of all living entities. It is not unexpected that efforts have already commenced to develop reliable and easy-to-implement techniques that will help establish the order of bases in DNA. The currently employed methods of such DNA sequencing continue to be complicated and time-consuming: they involve elaborate schemes of optical detection and computer-intensive sequence determination in order that information is obtained on the biochemistry that drives DNA fragmentation, amplification and chain termination. Next generation DNA sequencing (NGS) seeks to improve the throughput of the process, at lower costs [112, 113], but issues connected with errors and an apparent limit on the length of DNA that can be probed [114] continue to be barriers. Perhaps the use of tunnelling DNA sequencing might prove to be useful.

In nanopore sequencing [115–117] single-stranded DNA (in a physiologically-appropriate aqueous solution) is forced through a nanosized pore. As noted, femtosecond laser writing is employed to fabricate such pores. Ion-induced currents are generated as the nucleotides block the nanopore, and such currents can be correlated with the type of base in a given nucleotide. Nanopores used in early studies were made from biological materials [118, 119] but it is likely that graphene nanopores may be of utility in future. Graphene offers the advantage of providing an atomic layer thickness which might permit the attainment of single-base resolution [120–122]. The transverse ion current generated in such nanopores is, of course, due to tunnelling—an ultrafast process. Both the magnitude of the tunnelling current as well as the structural noise that accompanies it, carry signatures of the geometrical alignment of proximate bases. It has been possible to carry out modelling exercises by treating DNA as a large-bandgap amorphous semiconductor (with a π-π bandgap of $\sim$4 eV) [123].

11.9 Future Opportunities Offered by Ultrafast Biophotonics

Biological systems achieve several functionalities—respiration involving electron tunneling, photosynthesis involving quantum coherence, vision and DNA photolyases—with efficiencies that are usually impossible to imagine in artificial systems. As we have seen throughout this book, many, if not all, of these capabilities rely on ultrafast phenomena that occur on nanometric lengthscales and that rely on very precise arrangements of molecules. Of course, as with everything in nature, all such molecular arrangements are achieved through self assembly. Ultrafast biophotonics, along with methods of synthetic biology, offers exciting opportunities to develop deeper understanding of these biological functionalities as well as for making progress in mimicking nature's design principles in artificial devices. Potential areas for developing efficient biomimetic ultrafast devices involve solar energy harvesting, sensing, energy storage, and molecular electronics: many other possibilities await exploration.

References

1. W. Min, B.P. English, G. Luo, B.J. Cherayil, S.C. Kou, X.S. Xie, Acc. Chem. Res. **38**, 923 (2005)
2. D. Brinks, F.D. Stefani, F. Kulzer, R. Hildner, T.H. Taminiau et al., Nature **465**, 905 (2010)
3. R. Henderson, Quart. Rev. Biophys. **28**, 171 (1995)
4. R. Neutze, R. Wout, D. van der Spoel, E. Weckert, J. Hajdu, Nature **406**, 752 (2000)
5. H.N. Chapman, S.P. Hau-Riege, M.J. Bogan, S. Bajt, A. Barty et al., Nature **448**, 676 (2007)
6. H.N. Chapman, P. Fromme, A. Barty, T.A. White, R.A. Kirian et al., Nature **470**, 73 (2011)
7. M.M. Seibert, T. Ekeberg, F.R.N.C. Maia, M. Svenda, J. Andreasson et al., Nature **470**, 78 (2011)
8. B. La Scola, S. Audic, C. Robert, L. Jungang, X. de Lamballerie et al., Science **299**, 2033 (2003)
9. M.R. Sawaya, D. Cascio, M. Gingery, J. Rodriguez, L. Goldschmidt et al., Proc. Nat. Acad. Sci. **111**, 12769 (2014)
10. R. Koopmann, K. Cupelli, L. Redecke, K. Nass, D.P. Deponte et al., Nat. Methods **9**, 259 (2012)
11. Y. Takayama, S. Maki-Yonekura, T. Oroguchi, M. Nakasako, K. Yonekura, Sci. Rep. **5**, 08074 (2015)
12. M. Nakasako, Y. Takayama, T. Oriquichi, Y. Sekiquchi, A. Kobayashi et al., Rev. Sci. Instrum. **84**, 093705 (2013)
13. C. Rischel, A. Rousse, I. Uschmann, P.-A. Albouy, J.-P. Geindre et al., Nature **390**, 490 (1997)
14. J. Miao, P. Charalambous, J. Kirz, D. Sayre, Nature **400**, 342 (1999)
15. H. Ihee, V.A. Lobastov, U.M. Gomez, B.M. Goodson, R. Srinivasan et al., Science **291**, 458 (2001)
16. J.C.H. Spence, R.B. Doak, Phys. Rev. Lett. **92**, 198102 (2004)
17. W.E. Moerner, M. Orrit, Science **283**, 1670 (1999)
18. S. Weiss, Science **283**, 1676 (1999)
19. Y.C. Sasaki, Y. Okumura, S. Adachi, H. Suda, Y. Taniguchi, N. Yagi, Phys. Rev. Lett. **87**, 248102 (2001)
20. N. Ogawa, K. Hoshisashi, H. Sekiguchi, K. Ichiyanagi, Y. Matsushita et al., Sci Rep. **3**, 2201 (2013)
21. R.S. Mathew-Fenn, R. Das, P.A.B. Harbury, Science **322**, 446 (2008)
22. R.S. Mathew-Fenn, R. Das, J.A. Silverman, P.A. Walker, P.A.B. Harbury, PLoS One **3**, e3229 (2008)
23. X. Shi, D. Herschlag, P.A.B. Harbury, Proc. Natl. Acad. Sci. USA **110**, E1444 (2013)
24. D. Sayre, Acta Crystallogr. **5**, 843 (1952)
25. Y. Nishino, Y. Takahashi, N. Imamoto, T. Ishikawa, K. Maeshima, Phys. Rev. Lett. **102**, 018101 (2009)
26. M.A. Pfeifer, G.J. Williams, I.A. Vartanyants, R. Harder, I.K. Robinson, Nature **442**, 63 (2006)
27. J.R. McIntosh, *Cellular Electron Microscopy, Methods in Cell Biology* (Academic Press, San Diego, 2007)
28. J. Kirz, C. Jacobsen, M. Howells, Quart. Rev. Biophys. **28**, 33 (1995)
29. W. Chao, B.D. Harteneck, J.A. Liddle, E.H. Anderson, D.T. Attwood, Nature **435**, 1210 (2005)
30. J.R. Fienup, Appl. Opt. **21**, 2758 (1982)
31. M. Germann, T. Latychevskaia, C. Escher, H.-W. Fink, Phys. Rev. Lett. **104**, 095501 (2010)
32. M. Hoshino, K. Uesugi, N. Yagi, S. Mohri, J. Regini, B. Pierscionek, PLoS One **6**, e25140 (2011)
33. A. Momose, S. Kawamoto, I. Koyama, Y. Hamaishi, K. Takai, Y. Suzuki, Jpn. J. Appl. Phys. **42**, L866 (2003)
34. A. Momose, Jpn. J. Appl. Phys. **44**, 6355 (2005)
35. J. Wang, Nucleic Acids Res. **28**, 3011 (2000)
36. M.L. Larramendy, W. El-Rifai, S. Knuutila, Cytometry **31**, 174 (1998)

37. Z. Zhu, A.S. Waggoner, Cytometry **28**, 206 (1997)
38. X.C. Zhang, Phys. Med. Biol. **47**, 3667 (2002)
39. W. Zhang, E.R. Brown, M. Rahman, M.L. Norton, Appl. Phys. Lett. **102**, 023701 (2013)
40. T.R. Globus, D.L. Woolard, T. Khromova, T.W. Crowe, M. Bykhovskaia et al., J. Biol. Phys. **29**, 89 (2003)
41. A. Markelz, S. Whitmire, J. Hillebrecht, R. Birge, Phys. Med. Biol. **47**, 3797 (2002)
42. R.M. Woodward, V.P. Wallace, D.D. Arnone, E.H. Linfield, M. Pepper, J. Biol. Phys. **29**, 257 (2003)
43. S.E. Whitmire, D. Wolpert, A.G. Markelz, J.R. Hillebrecht, J. Galan, R.R. Birge, Biophys. J. **85**, 1269 (2003)
44. C.F. Zhang, E. Tarhan, A.K. Ramdas, A.M. Weiner, S.M. Durbin, J. Phys. Chem. B **108**, 10077 (2004)
45. M. Tang, Q. Huang, D. Wei, G. Zhao, T. Chang et al., J. Biomed. Opt. **20**, 095009 (2015)
46. A. Schliesser, N. Picqué, T.W. Hänsch, Nature Photon. **6**, 440 (2012)
47. Z. Movasaghi, S. Rehman, I. ur Rehman, Appl. Spectrosc. Rev. **43**, 134 (2008)
48. D.I. Ellis, R. Goodacre, Analyst **131**, 875 (2006)
49. T.W. Hänsch, Rev. Mod. Phys. **78**, 1297 (2006)
50. A. Schliesser, M. Brehm, F. Keilmann, D.W. van der Weide, Opt. Exp. **13**, 9029 (2005)
51. S.A. Diddams, L. Hollberg, V. Mbele, Nature **445**, 627 (2007)
52. I. Coddington, W.C. Swann, N.R. Newbury, Phys. Rev. Lett. **100**, 013902 (2008)
53. J. Mandon, G. Guelachvili, N. Picqué, Nature Photon. **3**, 99 (2009)
54. A. Foltynowicz, P. Masłowski, T. Ban, F. Adler, K.C. Cossel et al., Faraday Discuss. **150**, 23 (2011)
55. A.M. Zolot, F.R. Giorgetta, E. Baumann, J.W. Nicholson, W.C. Swann et al., Opt. Lett. **37**, 638 (2012)
56. A.A. Savchenkov, A.B. Matsko, D. Strekalov, M. Mohageg, V.S. Ilchenko, L. Maleki, Phys. Rev. Lett. **93**, 243905 (2004)
57. P. Del'Haye, A. Schliesser, O. Arcizet, T. Wilken, R. Holzwarth, T.J. Kippenberg, Nature **450**, 1214 (2007)
58. T. Udem, R. Holzwarth, T.W. Hänsch, Nature **416**, 233 (2002)
59. S.T. Cundiff, J. Ye, Rev. Mod. Phys. **75**, 325 (2003)
60. S.T. Cundiff, J. Ye, *Femtosecond Optical Frequency Comb: Principle, Operation and Applications* (Springer, New York, 2005)
61. B. Bernhardt, A. Ozawa, P. Jacquet, M. Jacquey, Y. Kobayashi et al., Nature Photon. **4**, 55 (2010)
62. M.J. Thorpe, K.D. Moll, R. Jason, Jones, B. Safdi, J. Ye, Science **311**, 1595 (2006)
63. M.J. Thorpe, J. Ye, Appl. Phys. B **91**, 397 (2008)
64. Y. Bao, X. Yi, Z. Li, Q. Chen, J. Li, X. Fan, X. Zhang, Light Sci. Appl. **4**, e300 (2015)
65. G. Millot, S. Pitois, M. Yan, T. Hovhannisyan, A. Bendahmane et al., Nature Photon. **10**, 27 (2016)
66. T. Ganz, M. Brehm, H.G. von Ribbeck, D.W. van der Weide, F. Keilmann, New J. Phys. **10**, 123007 (2008)
67. P.R. Griffiths, J.A. De Haseth, *Fourier Transform Infrared Spectroscopy*, 2nd edn. (Wiley, New York, 2007)
68. A. Foltynowicz, P. Maslowski, A.J. Fleisher, B.J. Bjork, J. Ye, App. Phys. B **110**, 163 (2013)
69. S. Amarie, F. Keilmann, Phys. Rev. B **83**, 045404 (2011)
70. E. Baumann, F.R. Giorgetta, W.C. Swann, A.M. Zolot, I. Coddington, N.R. Newbury, Phys. Rev. A **84**, 062513 (2011)
71. T. Ideguchi, S. Holzner, B. Bernhardt, G. Guelachvili, N. Picqué, T.W. Hänsch, Nature **502**, 355 (2013)
72. M. Brehm, A. Schliesser, F. Keilmann, Opt. Express **14**, 11222 (2006)
73. S.T. Cundiff, A.M. Weiner, Nature Photon. **4**, 760 (2010)
74. S.J. Lee, B. Widiyatmoko, M. Kourogi, M. Ohtsu, Jpn. J. Appl. Phys. **40**, L878 (2001)
75. D.J. David, S.A. Rice, J. Chem. Phys. **83**, 5013 (1985)

76. P. Brumer, M. Shapiro, Chem. Phys. Lett. **126**, 541 (1986)
77. R.J. Gordon, S.A. Rice, Ann. Rev. Phys. Chem. **48**, 601 (1997)
78. D. Mathur, F.A. Rajgara, J. Chem. Phys. **120**, 5616 (2004)
79. D. Mathur, K. Dota, A.K. Dharmadhikari, J.A. Dharmadhikari, Phys. Rev. Lett. **110**, 083602 (2013)
80. R. Bartels, S. Backus, E. Zeek, L. Misoguti, G. Vdovin et al., Nature **406**, 164 (2000)
81. N. Dudovich, D. Oron, Y. Silberberg, Nature **418**, 512 (2002)
82. R.S. Judson, H. Rabitz, Phys. Rev. Lett. **68**, 1500 (1992)
83. A.M. Weiner, Rev. Sci. Instrum. **71**, 1929 (2000)
84. K. Hitoshi, M.M. Wefers, K.A. Nelson, Annu. Rev. Phys. Chem. **46**, 627 (1995)
85. A. Assion, T. Baumert, M. Bergt, T. Brixner, B. Kiefer et al., Science **282**, 919 (1998)
86. B. Constantin, R. Chakrabarti, H. Rabitz, New J. Phys. **12**, 075008 (2010)
87. M. Dantus, V.V. Lozovoy, Chem. Rev. **104**, 1813 (2004)
88. D. Meshulach, Y. Silberberg, Nature **396**, 239 (1998)
89. Y. Silberberg, Annu. Rev. Phys. Chem. **60**, 277 (2009)
90. E. Knill, R. Laflamme, R. Martinez, C.-H. Tseng, Nature **404**, 368 (2000)
91. M. Viteau, A. Chotia, M. Allegrini, N. Bouloufa, O. Dulieu et al., Science **321**, 232 (2008)
92. C.-Y. Lien, C.M. Seck, Y.-W. Lin, J.H.V. Nguyen, D.A. Tabor, B.C. Odom, Nat. Commun. **5**, 4783 (2014)
93. M. Aeschlimann, T. Brixner, A. Fischer, C. Kramer, P. Melchior et al., Science **333**, 1723 (2011)
94. C.J. Bardeen, V.V. Yakovlev, J.A. Squier, K.R. Wilson, J. Am. Chem. Soc. **120**, 13023 (1998)
95. W. Wohlleben, T. Buckup, J.L. Herek, M. Motzkus, Chem. Phys. Chem **6**, 850 (2005)
96. H.A. Frank, A.J. Young, G. Britton, R.J. Cogdell, *The Photochemistry of Carotenoids* (Kluwer, Dordrecht, 1999)
97. V.I. Prokhorenko, A.M. Nagy, S.A. Waschuk, L.S. Brown, R.R. Birge, R.J.D. Miller, Science **313**, 1257 (2006)
98. D. Mathur, K. Dota, D. Dey, A.K. Tiwari, J.A. Dharmadhikari et al., J. Chem. Phys. **143**, 244310 (2015)
99. M.A. Taylor, J. Janousek, V. Daria, J. Knittel, B. Hage et al., Nature Photon. **7**, 229 (2013)
100. F. Wolfgramme, A. Cerè, F.A. Beduini, A. Predojević, M. Koschorreck, M.W. Mitchell, Phys. Rev. Lett. **105**, 053601 (2010)
101. J. Abadie et al., the LIGO Scientific Collaboration. Nature Phys. **7**, 962 (2011)
102. M.B. Nasr, D.P. Goode, N. Nguyen, G. Rong, L. Yang, B.M. Reinhard, B.E.A. Saleh, M.C. Teich, Opt. Commun. **282**, 1154 (2011)
103. A. Crespi, M. Lobino, J.C.F. Matthews, A. Politi, C.R. Neal et al., Appl. Phys. Lett. **100**, 233704 (2012)
104. G. Brida, M. Genovese, I.R. Berchera, Nature Photon. **4**, 227 (2010)
105. V. Giovannetti, S. Lloyd, L. Maccone, Science **306**, 1330 (2004)
106. P. Nandi, G. Jose, C. Jayakrishnan, S. Debbarma, K. Chalapathi et al., Opt. Express **14**, 12145 (2006)
107. J.A. Dharmadhikari, K. Pradyna, A. Bhatnagar, D. Mathur, A.K. Dharmadhikari, Opt. Commun. **287**, 122 (2013)
108. J.A. Dharmadhikari, R. Bernard, A.K. Bhatnagar, D. Mathur, A.K. Dharmadhikari, Opt. Lett. **38**, 172 (2013)
109. M. Garg, A.K. Tiwari, D. Mathur, J. Phys. Chem. A **116**, 8762 (2012)
110. D. Mathur, A.K. Tiwari, in *Concepts and Methods in Modern Theoretical Chemistry*, vol. 2, S. K. Ghosh, P. Chattaraj (Eds.) (Taylor and Francis, Boca Raton, 2013)
111. S.R. Leone, C.W. McCurdy, J. Burgdörfer, L.S. Cederbaum, Z. Chang et al., Nature Photon. **8**, 162 (2014)
112. J. Shendure, H. Ji, Nature Biotechnol. **26**, 1135 (2008)
113. M.L. Metzker, Nat. Rev. Genet. **11**, 31 (2009)
114. C.W. Fuller, L.R. Middendorf, S.A. Benner, G.M. Church, T. Harris et al., Nature Biotechnol. **27**, 1013 (2009)

115. M. Di Ventra, Nanotechnology **24**, 342501 (2013)
116. R.H. Scheicher, A. Grigoriev, R. Ahuja, J. Mater. Sci. **47**, 7439 (2012)
117. B.M. Venkatesan, R. Bashir, Nature Nanotech. **6**, 615 (2011)
118. J.J. Kasianowicz, E. Brandin, D. Branton, D.W. Deamer, Proc. Natl. Acad. Sci. USA **93**, 13770 (1996)
119. J. Li, D. Stein, C. McMullan, D. Branton, M.J. Aziz, J.A. Golovchenko, Nature **412**, 166 (2001)
120. K.K. Saha, M. Drndić, B.K. Nikolić, Nano Lett. **12**, 50 (2012)
121. T. Nelson, B. Zhang, O.V. Prezhdo, Nano Lett. **10**, 3237 (2010)
122. H.W.C. Postma, Nano Lett. **10**, 420 (2010)
123. J.R. Alvarez, D. Skachkov, S.E. Massey, J. Lu, A. Kalitsov, J.P. Velev, Phys. Rev. Appl. **1**, 034001 (2014)

Index

P. Vasa and D. Mathur, *Ultrafast Biophotonics*, Biological and Medical Physics, Biomedical Engineering, DOI 10.1007/978-3-319-39614-9

MIX
Papier aus verantwortungsvollen Quellen
Paper from responsible sources
FSC® C105338
FSC
www.fsc.org

If you have any concerns about our products,
you can contact us on
ProductSafety@springernature.com

In case Publisher is established outside the EU,
the EU authorized representative is:
Springer Nature Customer Service Center GmbH
Europaplatz 3, 69115 Heidelberg, Germany

Printed by Libri Plureos GmbH
in Hamburg, Germany